AF657523

HYDRAULIC MACHINES

(Fluid Machinery)

HYDRAULIC MACHINES

(Fluid Machinery)

R. K. Singal
Professor and Consultant
rk.singal@gmail.com

Mridul Singal
B.Tech. (IIT, Kanpur), M.Tech. (IISC, Bangalore)
I/T Architect, IBM India Pvt. Ltd.

Rishi Singal
B.Tech. (PEC Chandigarh), PMP (PMI, USA)
Accenture India Pvt. Ltd.

TECHSAR™
NEW DELHI

Published and Printed by

Techsar Pvt. Ltd.
4435-36/7, Ansari Road, Daryaganj
New Delhi–110 002 (India)
E-mail: tech_info@techsarworld.com
Website: www.techsarworld.com

ISBN: 978-93-80026-01-5

Reprint 2025

Preface

India is bestowed with large hydel power potential and can set up 1,50,000 MW capacity. Presently, only about 30,000 MW of capacity has been created. There is a big gap to exploit this renewable and clean source of energy. Every industrial process involves the transfer of liquid from one level of pressure to another. Pumps have become essential part of all industrial processes and, in turn, major consumers of energy. Therefore, hydraulic machines mainly consisting of hydraulic turbines and pumps have been included as a core subject for undergraduate and postgraduate students of Mechanical Engineering, Electrical Engineering, Civil Engineering, Hydraulic Engineering, Power Engineering, etc.

The present book has been designed to meet the above requirements. The book will be useful for the students, teachers and library reference. The requirements of engineers interested in taking up competitive examinations and research career have been fully taken care of.

Authors

Contents

PART II—TURBINES 43

PART I

FUNDAMENTALS

CHAPTER 1

Classification and Applications

1.1 INTRODUCTION

Water is by far the most common liquid on the earth and is therefore plentiful and cheap. It is also chemically stable and non-hazardous to health. H_2O molecules have extra strong mutual attraction due to hydrogen bond without which, it will be a gas at atmospheric temperature and life would not exist on the earth. It is also an extremely good solvent. At typical atmospheric conditions, density of water is 1000 kg/m^3. Hence, size and cost of plants using water as working medium are not impractically large.

Water is not common everywhere in the universe. It does not exist in stars because of break-up of molecules at very high temperatures. It does not exist on small planets because gravitational forces are too weak to trap the molecules. However, earth's planetary position, size and mean temperature are such as to contain and hold a more representative sample of water and it is therefore, plentiful and cheap on earth.

1.2 HYDRO ENERGY

Rain falling on the earth's surface has potential energy relative to sea level towards which it flows. This hydro energy can be converted into mechanical energy by shaft work where water falls through an appreciable vertical distance. The hydraulic power is thus a naturally available renewable energy source.

The utilisation of hydro energy to operate agricultural and industrial devices is one of the oldest and widespread techniques of human achievement. Hydraulic rams were used to raise water for lift irrigation and drinking purposes. Watermills have been used to generate mechanical power to drive irrigation pumps, machine tools and small electric generators. The Himalaya Mills is well known in the mountain areas from Afghanistan to Burma which is mostly used for grinding grains.

In modern hydroelectric power plants, the energy of water is utilised to drive a *turbine* which, in turn, runs the generator to produce electricity. A schematic diagram of a hydroelectric power plant is shown in Fig. 1.1.

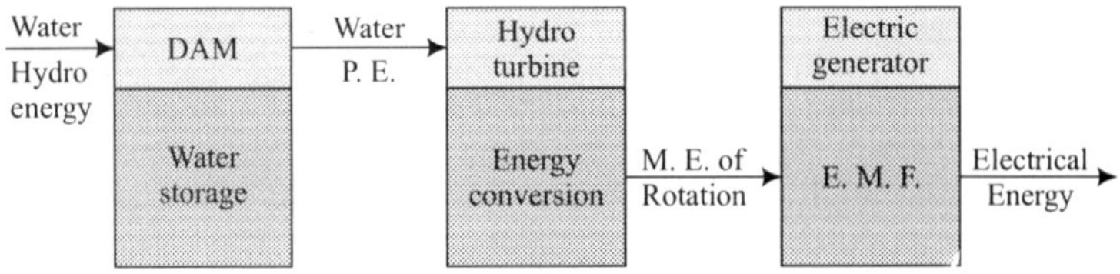

Fig. 1.1 Hydropower plant

The net electric power output of a hydroelectric power plant is given by:

$$P = \rho Q g H \eta_0$$

$$= \rho Q g H \, \eta_T \eta_g \eta_{Tr}$$

where P = Electric power output, [W]

ρ = Density of water, [1000 kg/m^3]

Q = Water flow rate in turbine, [m^3/s]

g = Gravitational acceleration, [9.81 m/s^2]

H = Net useful water head [m]

η_0 = Overall efficiency of hydropower plant

η_T = Turbine efficiency [0.85 – 0.95]

η_g = Generation efficiency [0.95 – 0.99]

η_{Tr} = Transformer efficiency [0.92 – 0.98]

Hydropower, available free of cost in the form of flowing water, is an important, pollution free, renewable source of energy. These can be integrated with irrigation, fish breeding and drinking water supply schemes. Nearly twenty per cent of the total world power is produced by hydel stations. There are some countries like Norway and Switzerland where the hydropower forms almost the total installed capacity.

The Central Electricty Authority (CEA) has assessed India's hydro potential to be about 1,48,700 MW of installed capacity. The hydroelectric capacity currently under operation is about 26,000 MW. Another 16,083 MW capacity is under various stages of development. India is endowed with economically viable hydro potential of 600 TWH.

The key advantage of hydroelectric power is the ability to store energy and the flexibility of its use during peak load periods.

Data of Central Electricity Authority shows that over a long period a hydel project is much cheaper than a thermal project. As hydel projects get older, their generator costs decrease, since fuel cost remains zero, while those of thermal projects would go up, on account of escalating fuel costs.

The water pumps especially centrifugal pumps are very widely used to transfer water from a lower level to higher level. Pumps are used in industries, municipal water supply, agriculture and other sectors.

1.3 CLASSIFICATION OF FLUID MACHINES

Liquids and gases together are called fluids. The fluid machines use liquids and/or gases as working substances for transfer and transformation of energy from one form to another. The fluid machines convert the potential energy or kinetic energy or pressure energy or thermal energy or internal energy of fluid into mechanical work and are called engines or turbines or prime movers or driving machines. The driven machines may be called pumps or compressors or blowers or fans to deliver fluids from low level of energy to higher level by consuming mechanical work. There are other fluid machines used for tranmission of energy from driver to driven machine such as a fluid coupling or torque converter. Fluid machines also include complete systems or components to intensify or store energy during transmission. There are so many types of machines working with different types of fluids, performing different functions and working on different principles that it may be difficult to classify them on single basis. These machines are grouped together on different criteria. A tree diagram is shown in Fig. 1.2 to broadly group different fluid machinery as per services provided by them.

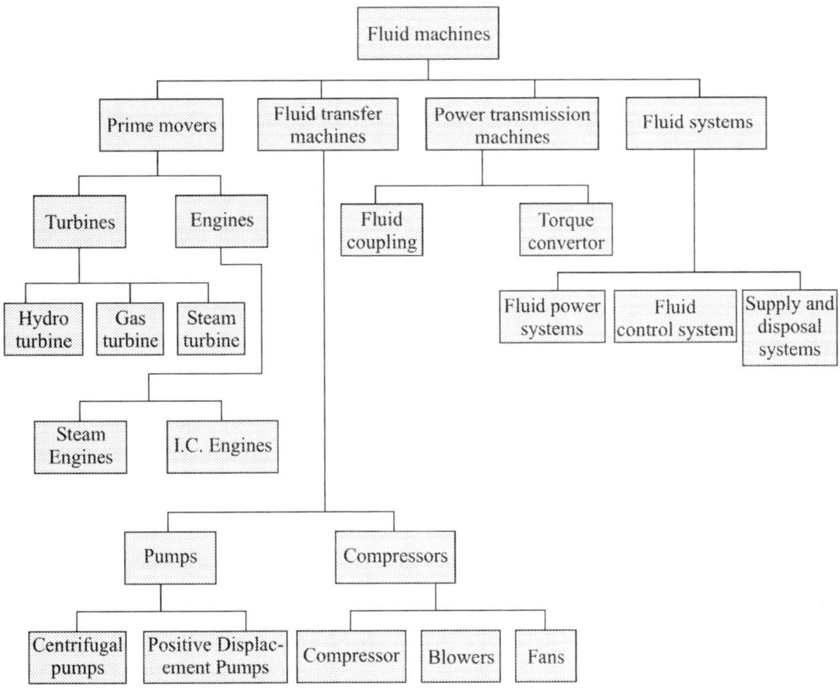

Fig. 1.2 Classification of fluid machines

1.3.1 Prime Movers

Prime movers are fluid machines which utilize the naturally available energy of fluid and transform into mechanical energy of shaft rotation. The natural sources of energy can be kinetic energy of wind, potential energy of water flow, solar heat, thermal energy produced by combustion of fuels (coal, gas, diesel, petrol, biomass) or geothermal energy from the core of the earth.

A steam engine is a reciprocating machine where enthalpy of steam is converted into mechanical (piston) work. Similarly, internal combustion engines are run on hot air obtained by combustion of diesel, petrol or gas and are also of piston cylinder reciprocating type. Steam turbine is a rotary machine which also converts the enthalpy drop of steam into mechanical work of shaft rotation. The steam may be generated in a boiler by evaporation of water by combustion gases or solar energy or geothermal steams or heat of nuclear fission. Gas turbines are also rotary machines run by hot air or carbon dioxide or helium or other gases heated by combustion of a fuel. All these machines are widely used for power generation or driving of process machines such as a pump, compressor, blower, or other machines. These prime movers are beyond the scope of this book.

Hydel turbines convert the potential energy of naturally available water into mechanical energy of shaft rotation which in turn is used to run an electric generator to produce electricity. These machines work on the principle of fluid dynamics and are known as rotodynamic machines. The rotating element of the turbine has a number of buckets or vanes or blades and is called a rotor. The fixed casing enclosing the rotor is called a stator. The turbine rotors are turned by the impulse of a waterjet or aerofoil shape of blades and are respectively called impulse turbines (Piston wheel) and Pelton reaction turbines (Francis and Kaplan turbine).

1.3.2 Fluid Transfer Machines

Fluid machines are widely used to transfer fluids from lower level or pressure to higher level or pressure. Machines transferring air or other gases with a pressure ratio upto 1.15 are called fans. Compressors are water cooled fluid machines producing pressure ratios more than 1.15. Compressors without cooling provision are called the blowers. The compressors, fans and blowers are run by electric motors or engines or other prime movers and are beyond the scope of this book.

The pumps handle liquids. The positive displacement pumps transfer the input mechanical energy of prime mover into pressure energy or kinetic energy of fluid with the help of a piston or gear teeth or a sliding gate. The use of the pump type depends upon the properties of the fluid to be pumped and also the quantity and pressure head of the system installation.

The centrifugal pumps are most widely used. The pump impeller is provided with vanes enclosed in a volute casing. As the impeller rotates by an electric motor or a fluid prime mover, the liquid is pushed by centrifugal force from the pump centre to its periphery through the volute casing. The kinetic energy generated by the pump impeller is converted into fluid pressure in the volute casing.

There are other pumps like jet pumps, pneumatic lift pumps, etc. which will be discussed separately.

1.3.3 Power Transmission Machines

In industrial practice, fluid machines are sometimes used to transmit mechanical energy from the motor shaft to the shaft of driven machines. The two shafts are connected through a coupling. Sometimes, it is required that the two shafts should not be connected mechanically. The shafts are connected hydraulically. The motor shaft drives a centrifugal pump to generate fluid pressure which is used to drive a fluid turbine mounted on the driven shaft. The pump and turbine are built into a single unit with a closed hydraulic unit.

The hydraulic power transmission systems are of two types:

1. *Fluid coupling* transmits power with same torque on the driving and driven shaft.
2. *Fluid torque converter* provides for torque multiplication with the same power on the driving and driven shafts. In addition to a pump and a hydraulic turbine as in the case of a fluid coupling, a reactionary number is incorporated to augment the torque.

1.3.4 Fluid Systems

A fluid system is a circuit in which forces and power are transmitted through a fluid. When air or gas is used as working fluid, the system is called fluidics which is beyond the scope of this book. The fluid systems discussed transmit forces or power by the hydraulic pressure of a liquid. There is no motion and no dynamic effect of liquid is used in the operation of the system.

1. The circuit diagram of a typical *fluid power system* is shown in Fig. 1.3.

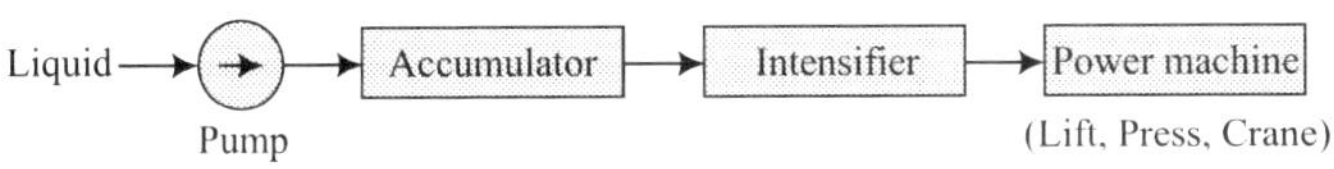

Fig. 1.3 Fluid system

The examples of fluid power systems are:

(i) The hydraulic press
(ii) The hydraulic lift
(iii) The hydraulic intensifier

(iv) The hydraulic accumulator

(v) The hydraulic crane

(vi) The hydraulic ram

2. The *fluid control systems* are used in nuclear power plants, thermal power plants, hydroelectric power plants and other process plants to operate and control the opening of water and steam valves; speed controller to regulate the load. Special servomotors are used to operate the valves.
3. The *supply and disposal systems* include lubrication system of turbines and generators, ash disposal system, effluent disposal system. Special slurry pumps, jet pumps and pneumatic lift pumps are used in the supply and disposal systems.

1.4 APPLICATIONS OF HYDRO TURBINES

The modern hydro turbines have been derived from the water wheels of the middle ages which were used for flour mills to grind wheat and ore crushing. Many technological developments in the field of fluid dynamics, metallurgy and mechanical engineering have helped to the development of very wide range and types of hydro turbines. There are turbines which can operate at heads of 200 m and wheel diameter of 10 m. Similarly, the power capacity of a turbine can be 700 MW and more. There are also turbine rotors derived from windmills which can work on zero head and utilize the current velocities of rivers and canals. Such water mills can be designed with water velocity of even 1m/s and power output of a few kilowatt. The main aim of all these turbines is to convert the potential energy of waterfalls or kinetic energy of flowing water into mechanical energy to produce electrical power.

The hydro turbines can be classified as:

1. High head (Pelton), medium head (Francis) or low head (Kaplan) turbines.
2. Hydro turbines for base load or peak load power plants.
3. Turbines for mini or micro hydel plants.
4. Turbines for hydro electric plants with storage reservoirs using a dam.
5. Turbines for run of the river plants without pondage.
6. Turbines for pumped storage plants.
7. Reversible turbines for tidal power plants.

The hydro power plants are designed as central power plants to generate electricity or as interconnected to thermal power plants in order to meet the fluctuating loads most economically.

Special reversible hydro turbines have been built for tidal power plants so that these can produce power both during high tide storage of water and low tide emptying of reservoir.

In the design of pumped storage plants, water from turbine discharge is stored in the tail race reservoir. When demand is low (at night), the water is pumped back from tail race to head water pond. The power required to drive the pump is taken from common grid or a nearby thermal power plant. During high demand (peak load) this pumped water is used to produce electricity. A motor generator set and a reversible turbine pump set can be used both for power generation and water pumping.

The small hydro plants can use the natural water source in hilly terrain for power generation. Zero head turbines can be used to tap the kinetic energy of river or canal water in plain areas.

The hydro electric power plants can be used for generating electricity only or as a multipurpose scheme for flood control, irrigation water and power generation.

1.4.1 Advantages of Hydel Power

Power generation by hydro turbine plants have the following attractions:

1. Hydro plants can be designed as multipurpose projects. In addition to generation of electricity, the project can be engineered for irrigation, flood control, afforestation, navigation and aquaculture.
2. These plants are simple in design and operation and have very high reliability and durability.
3. The hydraulic turbine can be started and stopped quickly in a few minutes. Therefore, no spinning or reserve capacity is needed.
4. The maintenance of plants is very simple and due to low temperature operation, life expectancy can be more than 50 years.
5. Hydro turbines have high efficiency (more than 90%) over a wide range of loads.
6. The running cost of a hydro turbine is very low due to zero cost of fuel, fuel transportation, low-skilled manpower and requirements. The cost of manpower and spare parts is very low.
7. The water after doing work in the turbine is still available for other applications such as drinking water schemes or irrigation of farms.
8. Water is a renewable and clean source of energy.
9. There is no adverse effect on the environment such as greenhouse effect and global warming.
10. There are no chimney gases containing oxides of sulphur or nitrogen or carbon.
11. There is no problem of ash disposal or processing of waste as from a nuclear power plant.

1.4.2 Disadvantages of Hydel Power

The main shortcomings are:

1. The initial cost of a hydro turbine plant is high due to huge and complex civil structures.
2. The hydro turbine has to be designed for every site as per head and run-off available. The cost and time of design and manufacture is high due to non-availability of standard products.
3. The plants are to be located near the source of water and suitable site of location. The cost of power transmission due to long transmission line and high power loss can be huge.
4. The gestation period of hydel projects may be ten to fifteen years affecting the cost and cost variability of the project.
5. Power plant design and operation depends upon the quantity of water available which varies over the year.
6. Large hydro plants using dams and large storages have very serious ecologically disturbing consequences. These can submerge large fertile areas destroy forests and vegetation and uproot people from villages affected. There is a strong opinion in favour of run-off river projects or small hydel projects against power projects with dams and reservoirs.

1.5 APPLICATIONS OF CENTRIFUGAL PUMPS

Every industrial process involves the transfer of liquids from one level of pressure or static energy to another. Pumps have become an essential part of all industrial processes and, in turn, major consumers of energy themselves. In many industries, pumps may form more than 40% of the auxillary equipment. The pumping capacity may be as high as 100,000 m^3/hr of liquid and heads as high as 2500 m of H_2O and more.

Centrifugal pumps are used in thermal power plants as boiler feed pumps, condenser circulating water pumps and feed water heater pumps. Such pumps are also used for fire protection system and other water services, Sealed centrifugal pumps are designed for nuclear power plants.

Jet pumps are used for evacuation of air from condensers. Rotary pumps are used in lubrication and control systems. Reciprocating pumps are used for dosing of chemicals into boilers. Centrifugal and jet pumps are designed for ash disposal systems.

Centrifugal pumps have maximum application. Some important applications of centrifugal pumps are given in Table 1.1.

Table 1.1 Applications of centrifugal pumps

S. No.	Sector	Activities
1.	*Rural sector*	
	• Agriculture	Irrigation, farming, insecticide and fertilizer spraying, forestry, fish farms and hatcheries
	• Water supply	Drinking and waste water
	• Cottage and rural industries	
2.	*Transport Sector*	
	• Harbours, docks and canals	
	• Road, rail and air transport	Transportation of equipment manufacture e.g. car, lorry, bus, aircraft, rolling stocks, naval
	• Marine	Commercial ship building, ship owners
	• Mining and quarrying	Coal mining, metallic and non-metallic ores, clay, salt, sound and gravel, etc.
3.	*Construction sector*	
	• Pipeline construction	
	• Civil engineering works	Harbours and dry docks, roads, bridges, etc.
4.	*Service industries and services*	
	• Heating, ventilating and air-conditioning	Hospitals, schools, offices, etc.
	• Domestic water supply	Domestic booster sets, hot water circulation
	• Fire-fighting and fire prevention	Sprinkler system, etc.
	• Armed forces, postal services, atomic energy establishment, communication, miscellaneous local government services	
	• Laundries, dry cleaners, dryers, sports establishments	
	• Commercial construction services	Hotels, hospitals, schools, offices
5.	*Power generation and distribution*	
	• Gas	Processing and distribution
	• Electricity	Electricity generation and transmission
	• Water supply	Municipalities and other agencies
	• Sewage	Sewage treatment and disposal
6.	*Oil exploration and production*	
	• Oil and natural gas	Exploration and production
	• Petroleum	Refinery plants, distribution centres
	• Coal, coke, tar	Processing
7.	*Engineering industry*	
	• Mechanical machinery and component manufacturers	Machine tools, pumps, compressors, food machinery, heaters, coolers, fastners, instruments, etc.

Table Contd.

Table Contd.

	• Boiler	Manufacturing
	• Dish and bottle washing plant	Manufacturing
	• Metal parts washing and spray painting	Manufacturing
	• Filtration plant	Manufacturing
	• Commercial industrial refrigeration	Manufacturing
	• Electrical motors, generators, radio, television, telephony equipment, wires and cable, other domestic electrical equipment	Manufacturing
	• Switchgear	Manufacturing
8.	*Food and beverages industry*	
	• Tobacco	
	• Cane and beet sugar processing	
	• Distilleries, breweries, soft drink	Manufacturing
	• Cereal processing, bakeries and confectionary products	
	• Canneries, fruit and vegetable processing and products	
	• Animal and marine fats and oils	
	• Dairy products, milk, cheese, butter, ice cream	
	• Meat processing, bacon production	Slaughter houses
9.	*Process industries*	
	• Petrochemicals, plastic and synthetic rubber	Manufacturing
	• Chemicals: fertilizers, inorganic and organic acid, solvents, ammonia, chlorine, inks, paints, phosphoric chemicals, polishers	Manufacturing
	• Water treatment	Plant manufacturers and contractors
	• Pharmaceutical and cosmetics	Manufacturing
	• Printing, publishing and book binding	
	• Paper, pulp and printing	Manufacturers and contractors
	• Textiles	Synthetic fibre manufacturers
	• Consumer goods	Textiles, tanneries, leather processing, furs, leather goods manufacturers, clothing and footwear, plastic consumer goods manufacturers
10.	*Industrial materials*	
	• Primary metal production, foundries, rolling mills	Equipment manufacturers and material processors
	• Electro-polishing and-plating	Plant manufacturers or process specialists
	• Cement ready mix concrete	
	• Bricks, fire clay, pottery glass and glass containers	Manufacturing
	• Timber, furniture, bedding and other wood and cork	Manufacturing

QUESTION BANK NO. 1

1. What is hydro energy? Why is water so common on earth's surface?
2. Describe the classification of fluid machines. What is the difference between fluid machines and hydel machines?
3. Explain the following:
   (a) Prime Movers
   (b) Fluid Transfer Machines
   (c) Power Transmission Machines
   (d) Fluid Systems
4. Briefly explain the applications of:
   (a) Fluid Systems.
   (b) Power Transmission Machines
5. What are the application of hydro turbines?
6. Discuss the advantages and disadvantages of hydel power.
7. Briefly explain the applications of centrifugal pumps.

CHAPTER 2

Basic Principles of Hydraulic Flow and Jet Theory

2.1 INTRODUCTION

Hydraulic machines handle water. Machines changing fluid energy into mechanical energy are called hydraulic turbines or hydraulic motors. Machines designed to more liquids and add energy to them are called pumps. There are also special systems dealing with the transmission of hydraulic power. The hydraulic machines are classified in Fig. 2.1.

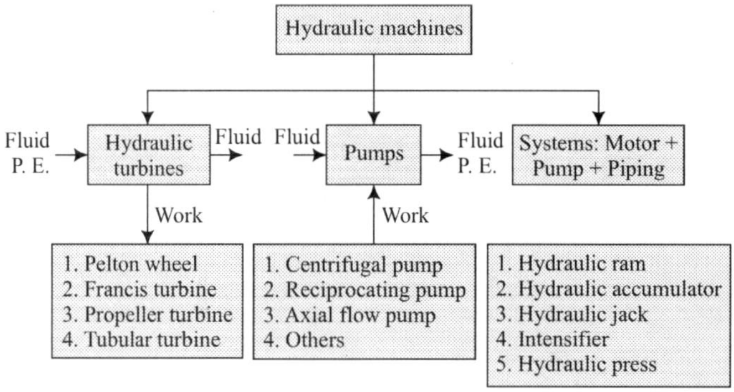

Fig. 2.1 Classification of hydraulic machines

The hydraulic mechanics are designed on the principles of fluid machanics, hydraulic flow and water jet forces.

2.2 PRINCIPLES OF FLUID MECHANICS

Important principles of fluid mechanics required for the design of fluid (hydraulic) machines are summarized below.

2.2.1 Hydrostatic Forces

1. Properties of fluids

The weight density or specific weight of a fluid is the weight per unit volume.

$$w = \rho \times g \quad [\text{N/m}^3]$$

Specific volume is the reciprocal of density.

$$\nu = \frac{1}{\rho} \quad [\text{m}^3/\text{kg}]$$

Kinematic υ is cosity,

$$\upsilon = \frac{\mu}{\rho} \quad [\text{stokes}]$$

2. Fluid pressure

The pressure at any point in a fluid is defined as the force per unit normal area.

$$p = \frac{F}{A} \quad [\text{N/m}^2]$$

The Pascal's law states that intensity of pressure for a fluid at rest is equal in all directions. Pressure variation at a point in a fluid at rest is given by the hydrostatic law which states that the rate of increase of pressure in the vertically downward direction is equal to the specific weight of fluid,

$$\frac{dp}{dz} = w = \rho \times g \quad [\text{N/m}^3]$$

The pressure at any point in a liquid

$$p = \rho g z \quad [\text{N/m}^2]$$

Absolute pressure = Atmospheric pressure + Gauge pressure

$$p_{abs} = p_{atm} + p_g$$

The pressure at a point in a fluid is measured by manometer.

3. Hydrostatic forces

The force exerted by a static fluid on a vertical, horizontal or an inclined immersed surface,

$$F = \rho g A \bar{h} \quad [\text{N}]$$

where $\bar{h}$ = depth of centre of gravity of the immersed surface from free surface of the liquid.

2.2.2 Kinematics of Fluid Flow

For a steady flow, the fluid properties like pressure, density, velocity, etc., do not change at a point with respect to time

$$\frac{dV}{dt} = 0, \quad \text{steady flow}$$

For uniform flow of fluid, the velocity does not change with respect to space (length of direction of flow)

$$\frac{dV}{ds} = 0, \quad \text{uniform flow}$$

The density of fluid remains constant for incompressible flow.

$$\rho = \text{constant: Incompressible flow.}$$

For laminar flow in a pipe, Reynolds number is less than 2000. For turbulent flow in a pipe, Reynolds number is more than 4000, Volumetric flow rate,

$$\dot{Q} = A_1 V_1 = A_2 V_2 = A_3 V_3 \quad [\text{m}^3/\text{s}]$$

This is called continuity equation.

2.2.3 Dynamics of Fluid Flow

1. Equation of motion (Newton's second law of motion)

$$F_x = m\, a_x \quad [\text{N}]$$

2. Euler's equation of motion along a stream line,

$$\frac{\partial p}{\rho} + g\partial z + V\partial V = 0$$

3. Bernaulli's equation (integration of Euler's equation of motion for a steady, ideal flow of an incompressible fluid), states that total energy consisting of pressure energy, kinetic energy and potential energy at any point of fluid is constant.

$$\frac{p_1}{\rho g} + \frac{V_1^2}{2g} + z_1 = \frac{p_2}{\rho g} + \frac{V_2^2}{2g} + z_2 + h_L \, [m]$$

where $\frac{p_1}{\rho g}$ = pressure head = pressure energy per unit weight,

$\frac{V_1^2}{2g}$ = kinetic head = kinetic energy per unit weight

z_1 = datum head = datum energy per unit weight

h_L = loss of energy head between sections 1 and 2

4. Momentum equation states that net force acting on a fluid mass is equal to change in momentum per second in that direction

$$F = \frac{d}{dt}(\mathrm{mV}) \quad [\mathrm{N}]$$

5. The impulse momentum equation is given as:

$$F\,dt = d\,(\mathrm{mV}) \quad [\mathrm{N\text{-}s}]$$

6. The force exerted by the nozzle on water

$$F_x = \rho Q(V_{2x} - \mathrm{V}_{1x}) \quad [\mathrm{N}]$$

7. Moment of momentum equation states that the resultant torque on a rotating fluid is equal to the rate of change of moment of momentum

$$T = \rho Q(V_2 r_2 - V_1 r_1) \quad [\mathrm{N\text{-}m}]$$

8. Loss of pressure head for viscous flow through circular pipe

$$h_f = \frac{32\mu \bar{u} L}{\rho g D^2} \quad [\mathrm{m}]$$

where μ = coefficient of viscosity

$\bar{u}$ = average fluid velocity

$$= \frac{Q}{\pi R^2} \ (\mathrm{m/s})$$

D = diameter of pipe [m].

9. *Darcy formula*. Energy loss due to friction,

$$h_f = \frac{4fLV^2}{2gD} \quad [\mathrm{m}]$$

10. *Darcy Weisbach Equation.* Head loss due to friction in pipes,

$$h_f = \frac{4fLV^2}{2gD} \quad [\mathrm{m}]$$

where f = Coefficient of friction

$$= \frac{16}{\mathrm{Re}} \text{ for laminar flow}$$

$$= \frac{0.0791}{(\mathrm{Re})^{1/4}} \text{ for turbulent flow.}$$

11. The velocity of water at the outlet of the nozzle,

$$V = \sqrt{\frac{2gH}{1 + \frac{4fL}{D} \cdot \frac{a^2}{A^2}}} \quad \text{[m/s]}$$

where H = head at inlet of pipe [m]

L = length of pipe [m]

D = diameter of pipe [m]

a = area of nozzle outlet [m^2]

A = area of pipe [m^2]

12. The power transmitted through nozzle,

$$P = \frac{\rho g Q}{1000}\left[H - \frac{4fLV^2}{2gN}\right] \quad \text{[kW]}$$

2.3 BASIC CONCEPTS OF HYDRAULIC FLOW

The hydraulic turbines utilize the potential energy of water to produce mechanical work. The pumps are required to transfer water and other liquids. The fluid couplings, torque converters and fluid system work with special oils.

2.3.1 Fluid Characteristics

The following fluid characteristics are assumed for the working substance used in hydraulic machines and systems.

1. Ideal fluid

An ideal fluid is an imaginary fluid which is both incompressible and non-viscous. Such liquids do not exist in nature. However, water is assumed to be incompressible and has very low value of viscosity. Therefore, it is nearly an ideal fluid. For incompressible fluid, the water density is constant.

$$\rho = \text{const.}$$

2. Newtonian fluid

The oils used in power transmission machines and fluid systems are also incompressible. These oils are assumed to be Newtonian fluids whose viscosity is independent of velocity gradient,

$$\tau = \mu \frac{du}{dy}$$

2.3.2 Closed and Open Systems

A system is a quantity of matter in space upon which attention is made in the study of changes of properties and analysis of a problem. Everything external to the system is called the surrounding. The boundary separating the system from the surrounding may be real (solid) or imaginary.

There may be energy transfer into or out of the system.

1. *Closed system.* A closed system has fixed identity with fixed mass. There is no mass (fluid) transfer across the system boundary.

   All positive displacement machines such as reciprocating pumps, gear pumps, etc. torque converters, fluid couplings, various fluid power systems and fluid control systems will be analyzed as constant mass closed systems.

2. *Open system.* In an open system, the fluid crosses the boundary of the system in addition to interaction of energy between the system and the surrounding. The mass of an open system may or may not change. The identity of the fluid changes continuously. The boundary of the open system is kept fixed without any change in its volume. An open system is also referred to as *control volume system*. The closed boundary of a control volume is called the *control surface*. There is a transfer of both mass and energy across the control surface. All types of hydraulic turbines, centrifugal pumps, axial flow pumps, slurry pumps, jet pumps, pneumatic lift pumps, various supply and disposal fluid systems will be analyzed and studied as control volume systems.

2.3.3 Macroscopic Properties

Matter is composed of several molecules and description of the motion of a fluid will consider the behaviour of discrete molecules which constitute the fluid. The microscopic properties can be found out by statistical summation of properties of individual molecules.

Liquids have extra-strong intermolecular attractive forces. Therefore, molecular description is not required. The entire liquid mass behaves as a continuous mass. The time-averaged values of properties are sufficiently accurate and valid. The matter of the system is assumed as a continuous distribution of mass with no empty space and no conglomeration of separate molecules. The fluid machines and system will be designed and analyzed on the basis of time-averaged macroscopic properties measured with the help of instruments. This is called the concept of *continuum* model.

2.3.4 Conservation of Mass

In non-nuclear processes, the matter can neither be created nor destroyed. The conservation of mass is inherent to the concept of a closed system. For a control volume, the rate of mass entering the system must be equal to the rate of mass leaving the system plus the rate of storage of mass in the system. If the fluid flow is steady, the rate of liquid stored is zero.

Most of the channels comprising hydraulic machines have variable and irregular sections and a curved mean path. Some of the channels are in circular motion with power applied to or taken from the flow.

The behaviour of hydraulic flow is studied making the following assumptions:

1. The flow is steady. The fluid properties do not change with time at any particular section.
2. The flow is incompressible and fluid density (ρ) remains constant.
3. The flow is one-dimensional.
4. The fluid velocity is uniform across the cross-section of the channel. Average velocity is assumed for each section of flow.
5. There is no branching of fluid stream.

2.3.5 Fluid Motion

A fluid mass consists of a number of fluid particles. The instantaneous velocity at any point in the fluid is the velocity of a particle that exists at that point at that time. The fluid motion can be studied by the following methods.

1. Langrangian method

A single fluid particle is followed during its motion and its velocity, acceleration, density, etc. are described. It is possible to trace the history of each fluid particle. However, it is very difficult to find solutions mathematically. The method is rarely used for practical applications.

2. Eulerian method

The velocity, acceleration, pressure, density, etc. are described at a certain point in flow field. It avoids the determination of movement of each individual particles of fluid. The change of velocity with time can be easily found out. The Eulerian method is commonly used in the analysis of fluid mechanics and fluid machines.

3. Stream line and stream tube

The motion of fluid through hydraulic machines can be described with the help of streamlines and stream tubes. *Streamline* is the instantaneous locus of a line in the flow field to which the velocity vector is tangent. The velocity at right angles to the streamline is always zero and no flow can cross any streamline.

Stream tube is a tubular space formed by the collection of streamlines. No fluid can enter or leave the stream tube except at the ends. Stream tube is imagined to behave as a solid tube. The cross-sectional areas of this tube are small in comparison with its length, so that velocity can be assumed constant across any cross-sectional area (A_i).

For steady flow, the stream tube maintains the same shape for all times while for unsteady flow, the stream tube changes shape with each succeeding instant of time.

2.3.6 Governing Equation

The flow of liquid through hydraulic machines is divided in imaginary stream tubes and the behaviour of the hydraulic flow is studied with the help of the following governing equations.

1. *Continuity Equation*

   This is based on the principle of conservation of mass flow ($\dot{m}$).

2. *Energy Equation*

   This is derived from the principle of conservation of energy ($\dot{m}$ V^2)

3. *Equation of Motion*

   This is based on the principle of conservation of momentum ($\dot{m}$ V).

2.4 CONTINUITY EQUATION

The continuity equation of flow is derived on the principle of conservation of mass. For an incompressible, uniform and steady state flow, the quantity of fluid entering at one end of stream tube and leaving at the other must be same provided there is no sink or source in the tube.

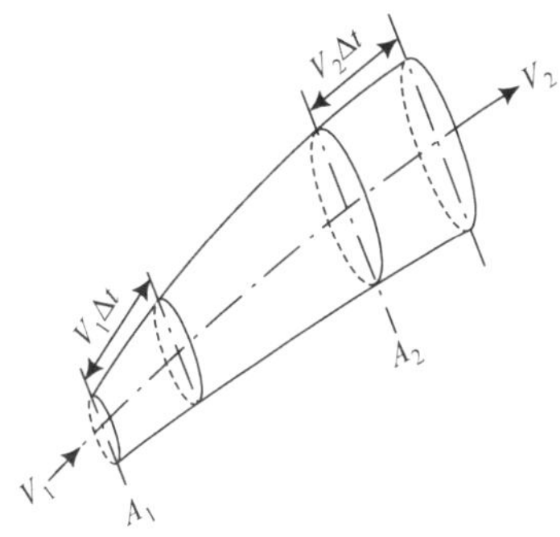

Fig. 2.2 Flow through a stream tube

Consider a stream tube with two arbitrary cross-sections A_1 and A_2, the mass of fluid enters the tube through A_1 with a velocity V_1 and leaves through A_2 with velocity V_2. In an increment of time Δt, the fluid particles entering A_1 have moved an infinitesimal distance $V_1\Delta t$. The particle at A_2 have moved a distance $V_2\Delta t$. By the definition of a stream tube no fluid can cross the stream tube. The mass of fluid entering the stream tube at A_1 in time Δt must be exactly equal to the mass of the fluid leaving the stream tube at A_2 during the same interval of time. For a steady flow, the rate of fluid stored is zero. Mathematically, the statement can be expressed as follows:

$$\dot{m}_1 = \dot{m}_2$$

$$\rho\dot{Q}_1 = \rho\dot{Q}_2$$

$$\rho A_1 V_1 \Delta t = \rho A_2 V_2 \Delta t$$

For incompressible flow, density ρ is constant.

$$\therefore \quad V_1 A_1 = V_2 A_2 = VA = \text{Constant} = \dot{Q} \quad [\text{m}^3/\text{s}]$$

This is called equation of continuity and the constant $\dot{Q}$ represents the volumetric flow rate. Therefore, the volume of fluid which passes through each cross-section of the stream tube per unit time remains constant for a steady flow and incompressible fluid.

2.5 THREE-DIMENSIONAL FLOW

In a three-dimensional fluid flow, the fluid properties (velocity, pressure, density, viscosity) may vary in all directions.

Examples are: flow in a river, flow within a fluid machine. However, for simplicity, the fluid machines are analyzed as one-dimensional fluid flow.

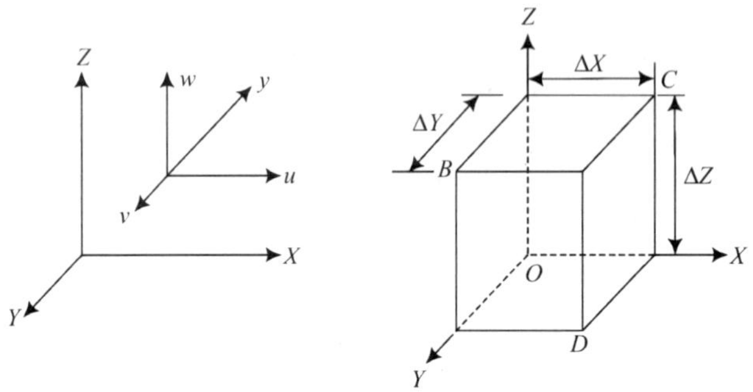

Fig. 2.3 Three-dimensional flow

Consider an elementary cube Δx, Δy, Δz in the fluid body. The difference between the amounts of fluid which flows into and out of its faces during time Δt, must be equal to the increase in the mass which the edges enclosed.

The mass of fluid of density ρ entering the cross-section across the face *OB* in time $\Delta t = \rho u(\Delta y.\Delta z)\Delta t$.

The mass of fluid leaving across the face *CD* in time $\Delta t = \rho u(\Delta y.\Delta z)\Delta t + \frac{\partial}{\partial x}(\rho u \Delta y \Delta z \Delta t)\Delta x$

$\therefore$ Gain of mass across the above faces $= -\frac{\partial}{\partial x}\rho u(\Delta x.\Delta y.\Delta z)\Delta t$

Similarly, gain of mass across the faces *BD* and $OC = -\frac{\partial}{\partial y}\rho v(\Delta x.\Delta y.\Delta z)\Delta t$

And, gain of mass across the faces *BC* and $OD = -\frac{\partial}{\partial z}\rho w(\Delta x.\Delta y.\Delta z)\Delta t$

$$\text{Total gain} = -\left(\frac{\partial}{\partial x}\rho u + \frac{\partial}{\partial y}\rho v + \frac{\partial}{\partial z}\rho w\right)(\Delta x.\Delta y.\Delta z)\Delta t. \qquad ...(1)$$

Mass of fluid contained at time $t = \rho(\Delta x.\Delta y.\Delta z)$

Mass of fluid contained at time $t + \Delta t = \rho(\Delta x.\Delta y.\Delta z) + \frac{\partial}{\partial t}(\rho.\Delta x.\Delta y.\Delta z)\Delta t$

$\therefore$ Gain in the mass of fluid in time $\Delta t = \dfrac{\partial}{\partial t}\rho.\Delta t(\Delta x.\ \Delta y.\Delta z)$ ...(2)

Equating equations (1) and (2),

$$-\left(\frac{\partial}{\partial x}\rho u + \frac{\partial}{\partial y}\rho v + \frac{\partial}{\partial z}\rho w\right)(\Delta x.\Delta y.\Delta z)\,\Delta t = \frac{\partial}{\partial t}\rho\Delta t(\Delta x.\Delta y.\Delta z)$$

or,
$$\frac{\partial \rho}{\partial t} + \frac{\partial}{\partial x}(\rho u) + \frac{\partial}{\partial y}(\rho u) + \frac{\partial}{\partial z}(\rho w) = 0 \quad \text{General Equation}$$

For incompressible fluid, $\rho = \text{Constant}$

$$\therefore \quad \frac{\partial u}{\partial x} + \frac{\partial v}{\partial y} + \frac{\partial w}{\partial z} = 0$$

This is also called differential equation of continuity in Cartesian coordinates.

Special Cases

1. Two-dimensional flow.

$$w = 0$$

$$\therefore \quad \frac{\partial u}{\partial x} + \frac{\partial v}{\partial y} = 0$$

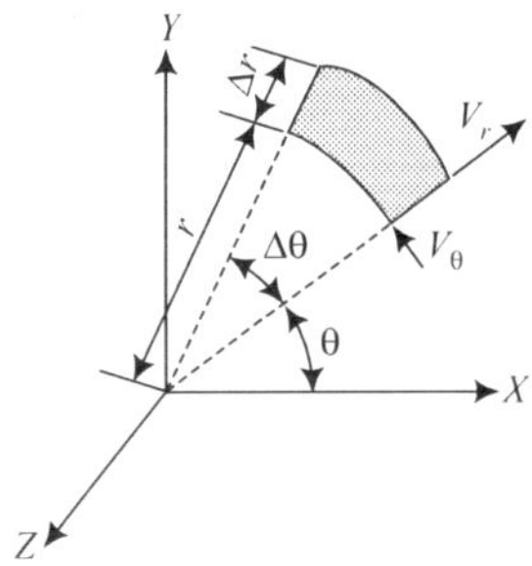

Fig. 2.4 Cylindrical coordinates

2. Cylindrical coordinates

$$\frac{\partial v_r}{\partial r} + \frac{1}{r}\frac{\partial v_\theta}{\partial \theta} + \frac{\partial v_z}{\partial z} + \frac{\rho v_r}{r} = 0$$

3. Spherical coordinates

$$\frac{1}{w}\frac{\partial}{\partial w}(w^2 v_w) + \frac{1}{\sin\phi}\frac{\partial(v_\phi \sin\phi)}{\partial \phi} + \frac{1}{\sin\phi}\frac{\partial v_\theta}{\partial \theta} = 0$$

4. Vector Notation

   General equation,

$$\frac{\partial \rho}{\partial t} + \nabla(\rho q) = 0$$

   For incompressible fluid,

$$\Delta q = 0$$

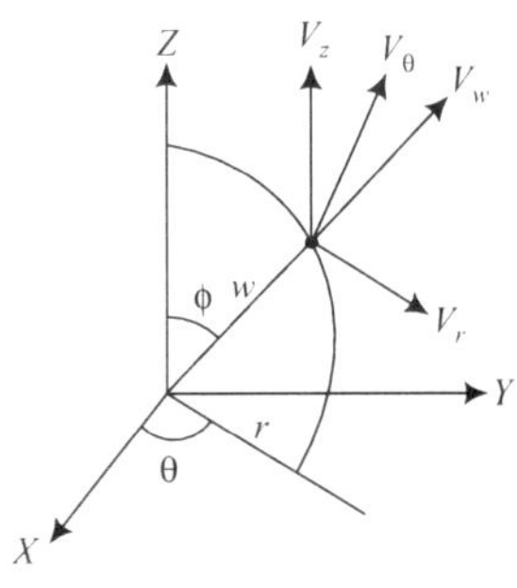

Fig. 2.5 Spherical coordinates

2.6 MOMENTUM EQUATION

The momentum equation is based on Newton's second law of motion which states that the time rate of change of momentum is proportional to the applied force and takes place in the direction of force.

Assume a resultant force F_x acts on a mass m along x-axis. The change in velocity of mass m in time dt be dV.

$$\therefore \quad F_X = m.\frac{du}{dt} \quad ...(3)$$

or

$$F_X dt = m\, du \quad ...(4)$$

Equation (3) is called linear momentum equation. Equation (4) is called impulse-momentum equation which shows that impulse of applied force ($F_X.dt$) is equal to change of momentum ($m.du$).

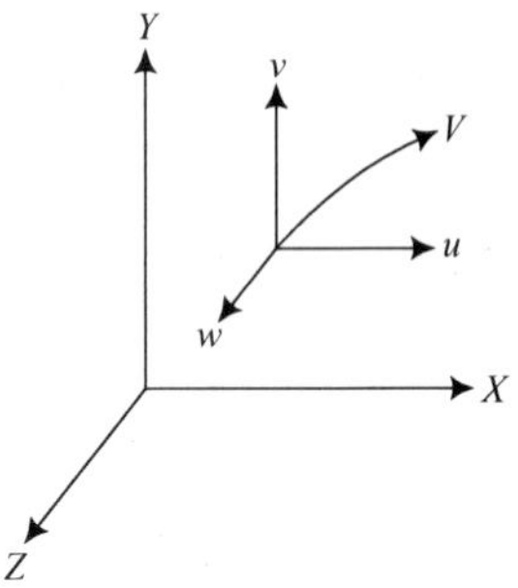

Fig. 2.6(a) Velocity components

Consider a stream tube. For steady state flow, mass entering the tube is equal to mass leaving the tube.

$$\rho_1 A_1 V_1 \Delta t = \rho_2 A_2 V_2 \Delta t$$

Momentum of fluid entering

$$= (\rho_1 A_1 V_1 \Delta t)\, V_1$$

Momentum of fluid leaving

$$= (\rho_2 A_2 V_2 \Delta t)\, V_2$$

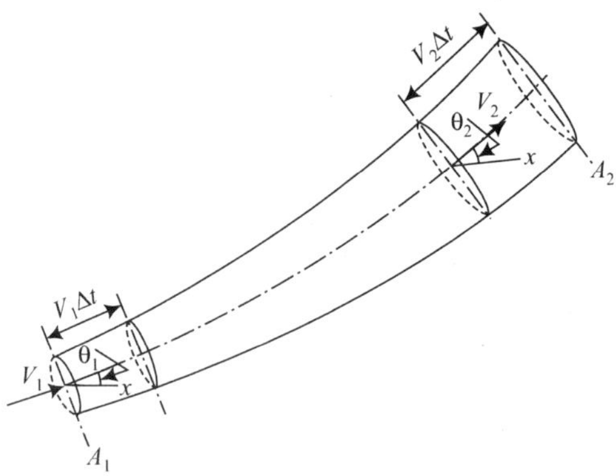

Fig. 2.6(b) Momentum equation

$\therefore$ Change of momentum in x-direction,

$$= [(\rho_2 A_2 V_2 \Delta t)\, V_2 \cos \theta_2] - [(\rho_1 A_1 V_1 \Delta t)\, V_1 \cos \theta_1]$$

But for a steady state, equation of continuity:

$$(\rho_1 A_1 V_1 \Delta t) = (\rho_2 A_2 V_2 \Delta t) = \rho Q \Delta t$$

$\therefore$ Change of momentum in x-direction,

$$= \rho Q \Delta t\,(V_2 \cos \theta_2 - V_1 \cos \theta_1)$$

From impulse-momentum equation,

$$\sum F_x \Delta t = \sum \rho Q \Delta t (V_2 \cos \theta_2 - V_1 \cos \theta_1)$$

$$F_x = \rho Q(V_2 \cos \theta_2 - V_1 \cos \theta_1) = \rho Q(v_{x_2} - v_{x_1}]$$

Similarly, $$F_y = \rho Q(V_2 \sin \theta_2 - V_1 \sin \theta_1) = \rho Q(v_{y_2} - v_{y_1})$$

Resultant force acting on the flowing fluid,

$$F = \sqrt{F_x^2 + F_y^2}$$

The angle between F and F_x can be found out as:

$$\tan \theta = \frac{F_y}{F_x}$$

The flowing fluid will also exert an equal and opposite force on the boundary of the stream tube. The forces may include dynamic force due to change of momentum, static pressure, weight of fluid, drag force due to friction, gravity and inertia forces due to centrifugal force.

2.7 APPLICATIONS OF MOMENTUM EQUATION

There are two types of applications of Impulse-Momentum equation.

1. Determination of forces exerted by the flowing fluid on the boundaries of flow passage of hydraulic machines due to change of momentum. The examples are:
   (i) Forces caused by a fluid jet striking a surface, i.e., fixed and moving blades of hydraulic machines.
   (ii) Jet propulsion.
   Propulsion of ships, boats, rockets, turbojet, ramjet, etc.
   (iii) Propellers.
   Marine propellers, helicopters.
   (iv) Pipe bends and reducers
2. Determination of flow characteristics due to energy loss in the flow systems. The examples are:
   (i) Sudden enlargement or constriction in a pipe such as orifices and mouth pieces.
   (ii) Hydraulic jump in open channel.

2.8 MOMENT OF MOMENTUM EQUATION

This equation is derived from the principle of moment of momentum which states that the resulting torque acting on a rotating fluid is equal to the rate of change of moment of momentum.

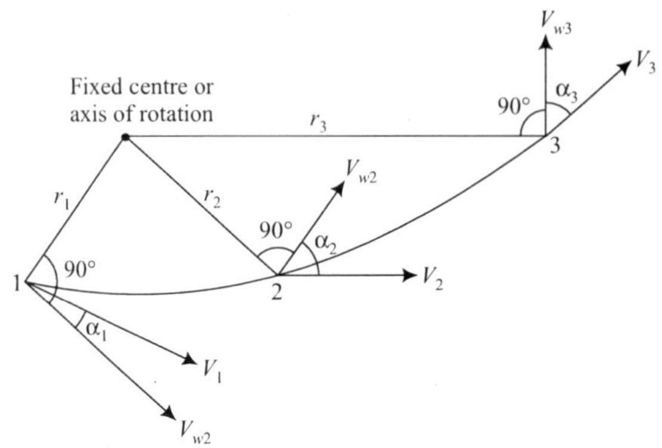

Fig. 2.7 Angular momentum

Let V_1 = velocity of fluid at section 1 of stream tube

r_1 = radius of curvature at section 1

ρ = density of fluid

Q = rate of flow of fluid

Momentum of a particle (mass m) in a direction normal to radius r

$$= m V_w$$

Moment of momentum of particle about axis of rotation

$$= m V_w r$$

From geometry,

$$V_w = V \cos \alpha$$

∴ Moment of momentum $= mV \cos \alpha r$

This is called angular momentum.

Angular momentum at section 1 $= mV_1 \cos \alpha_1 r_1$

Angular momentum at section 2 $= mV_2 \cos \alpha_2 r_2$.

∴ Change of angular momentum $= m(V_2 \cos \alpha_2 r_2 - V_1 \cos \alpha_1 r_1)$

Rate of change of angular momentum

$$= \frac{m}{t}(V_2 \cos \alpha_2 r_2 - V_1 \cos \alpha_1 r_1)$$

But rate of change of momentum is equal to torque.

$$\therefore \quad T = \frac{m}{t}(V_2 \cos \alpha_2 r_2 - V_1 \cos \alpha_1 r_1)$$

$$= \rho Q(V_2 \cos \alpha_2 r_2 - V_1 \cos \alpha_1 r_1)$$

For a circular path, $r_1 = r_2 = r_3$ and $\alpha_1 = \alpha_2 = \alpha_3 = 0$

$$\therefore \quad T = \rho Q r(V_2 - V_1)$$

Power $\quad P = Tw = \rho Q r(V_2 - V_1)w$

where w = angular velocity.

Blade velocity, $\quad V_b = wr$

$$\therefore \quad P = \rho Q(V_{w_2} V_{b_2} - V_{w_1} + V_{b_1})$$

The moment of momentum equation is applied for:

1. Analysis of flow problems in turbines and centrifugal pumps.
2. Finding torque exerted by water on sprinkler.

2.9 EULER'S FUNDAMENTAL EQUATION

Euler's fundamental equation is the equation of motion for a fluid with the following assumptions.

Assumptions

1. Fluid is non-viscous and frictional losses are zero.
2. Fluid is homogeneous and incompressible. ρ = constant
3. Flow is steady.
4. Flow is one-dimensional along the streamline.
5. Velocity is uniform over the section
6. Flow is continuous
7. Except gravity and pressure forces, no other forces are involved. Other forces like forces due to viscosity, turbulence and compressibility are neglected.

Energy is defined as ability to do work. It manifests in various forms and can change from one form to another.

1. Gravitational potential energy or elevation energy Z in metres of liquid column.
2. Kinetic energy due to mass and velocity of fluid $\frac{V^2}{2}$ in N-m.
3. Pressure energy required to move the fluid against its pressure.

   Pressure energy = p [N/m^2].

Consider a stream tube. The forces acting on a cylindrical element of stream tube are:

1. Pressure force pdA in the direction of flow
2. Pressure force $\left(p + \frac{\partial p}{\partial s} ds\right) dA$ opposite to the direction of flow.
3. Weight of the element, $\rho g\, dA\, dS$.

The resultant force on fluid element in the direction of S = mass × acceleration

$$\therefore \quad pdA - \left(p + \frac{\partial p}{\partial s}\right) dA - \rho g\, dA\, dS \cos\theta = \rho dA\, dS \times a_S \qquad ...(5)$$

where a_S = acceleration in the direction of S.

$$= \frac{dV}{dt}$$

$$V = V(S, t)$$

$$\therefore \quad a_S = \frac{\partial V}{\partial S} \cdot \frac{\partial S}{\partial t} + \frac{\partial V}{\partial t}$$

$$= V\frac{\partial V}{\partial S} + \frac{\partial V}{\partial t} \quad \left[\because \frac{dS}{dt} = V\right]$$

For steady flow, $\frac{\partial V}{\partial t} = 0$

$$\therefore \quad a_S = \frac{V\partial V}{\partial S}$$

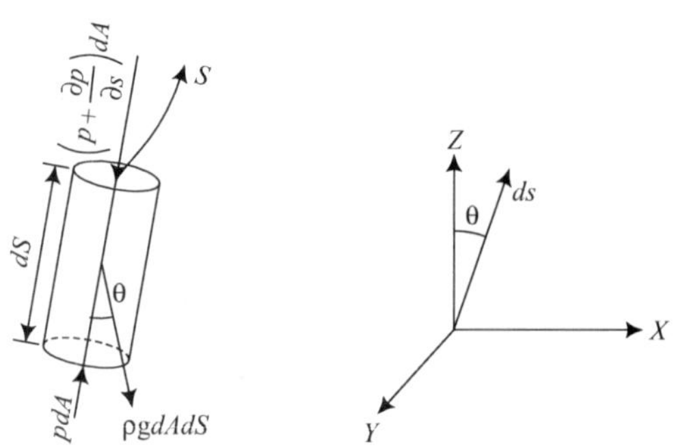

Fig. 2.8 Forces on a fluid element

From equation (5),

$$-\frac{\partial p}{\partial s} dA\, dS - \rho g\, dA\, dS \cos\theta = \rho dA\, dS \frac{V \partial V}{\partial S}.$$

Dividing throughout by $\rho dS\, dA$

$$\therefore \qquad -\frac{\partial p}{\rho \partial S} - g \cos\theta + V\frac{\partial V}{\partial S} = 0$$

But $$\cos\theta = \frac{dZ}{dS}$$

$$\therefore \qquad \frac{1}{\rho}\frac{\partial p}{\partial S} + g\frac{dZ}{dS} + V\frac{dV}{dS} = 0$$

or $$\frac{\partial p}{\rho} + g dZ + V dV = 0$$

or $$\frac{\partial p}{\rho} + g dZ + V dV = 0 \qquad \text{...(6)}$$

Equation (6) is called Euler's equation of motion.

2.9.1 Bernoulli's Equation

Bernoulli's equation is also called energy equation and is obtained by integration of Euler's equation of motion

$$\int \frac{dp}{\rho} + \int g dZ + \int V dV = \text{constant}$$

For incompressible flow, ρ = constant

$$\therefore \qquad \frac{p}{\rho} + gZ + \frac{V^2}{2} = \text{constant}$$

or $$\frac{p}{\rho g} + Z + \frac{V^2}{2g} = \text{constant} \qquad \text{...(7)}$$

Equation (7) is called Bernoulli's equation.

$\frac{p}{\rho g}$ = Pressure head = Pressure energy per unit weight of fluid.

$\frac{V^2}{2g}$ = Kinetic head = Kinetic energy per unit weight

Z = Potential head = Potential energy per unit weight.

The Bernoulli's equation can also be written as

$$\frac{p_1}{\rho g} + z_1 + \frac{V_1^2}{2g} = \frac{p_2}{\rho g} + z_2 + \frac{V_2^2}{2g}.$$

2.10 JET THEORY

When a nozzle is fitted at the end of a pipe, a jet of liquid comes out with high velocity utilizing the pressure of liquid in the pipe. Applying continuity equation to the sections 1 and 2 of the nozzle,

$$A_1 V_1 = A_2 V_2$$

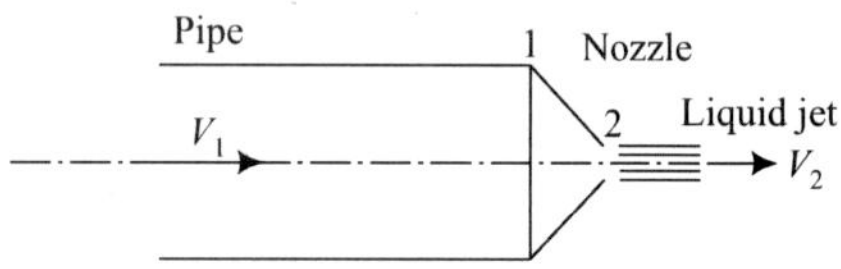

Fig. 2.9 Liquid jet

The velocity of liquid jet,

$$V_2 = \frac{A_1}{A_2} V_1$$

The kinetic energy of the jet,

$$KE_j = \frac{1}{2} \dot{m} V_2^2$$

Let jet velocity = $V_j = V_2$

$$\therefore \quad KE_j = \frac{1}{2} \dot{m} V_j^2$$

$$= \frac{1}{2} \rho A_j V_j \times V_j^2 = \frac{1}{2} \rho A_j V_j^3 \quad [j/s]$$

The jet power,

$$P_j = \frac{\frac{1}{2} \rho A_j V_j^3}{1000} \quad [\text{kW}]$$

The efficiency of nozzle,

$$\eta_n = \frac{\text{Jet power at outlet of nozzle}}{\text{Liquid power at inlet of nozzle}}$$

$$= \frac{\frac{1}{2 \times 1000} \rho A_j V_j^3}{\frac{\rho g \dot{Q} H}{1000}} = \frac{V_j^2}{2gH}$$

where H = total head of liquid at the inlet of nozzle.

The liquid jet exerts a force on a plate placed in front of it.

2.10.1 The Impulse Principle

The horizontal fluid jet impinging on a fixed vertical flat plate will spread out along the plate. The velocity in the direction of the jet is reduced to zero. A horizontal force F is imparted to the plate in the direction of the jet. This force is called an impulse and is equal to the change in momentum of the jet.

$$F = \dot{m}(V_j - 0) = \dot{m}\,V_j.$$

This principle is used in impulse turbines like Pelton wheel to get mechanical power of rotation from the jet force.

2.10.2 Reaction Principle

The fluid coming out of a fixed nozzle has a force which can be used to rotate a turbine wheel. There will be a corresponding and equal force tending to move the nozzle in the opposite direction. This opposite force is called reaction. If the nozzle is fitted on a wheel, the same will rotate in the opposite direction of the fluid jet and mechanical work of wheel rotation is produced.

The reaction principle is used in the reaction turbines i.e., Francis and Kaplan turbines to generate mechanical power of shaft rotation from the reaction of jet.

The force of reaction $= -\rho A V_j^2 = -\dot{m}\,V_j$,

2.10.3 Propulsion Principle

The reaction principle is also utilized in the propulsion of ships, rockets, vessels, jet aircraft, etc. The machine is fitted with a nozzle and is free to move in the direction opposite to that of jet. This is called propulsion principle or jet propulsion,

2.11 JET FORCE ON STATIONARY FLAT PLATE

2.11.1 Normal Plate

The initial velocity of jet $= V_j$

The final velocity of jet $= 0$

The impinging force,

$$F = \dot{m}(V_{jf} - V_{ji})$$
$$= \dot{m}(0 - V_j)$$
$$= -\dot{m}\,V_j = -\rho \dot{Q} V_j.$$

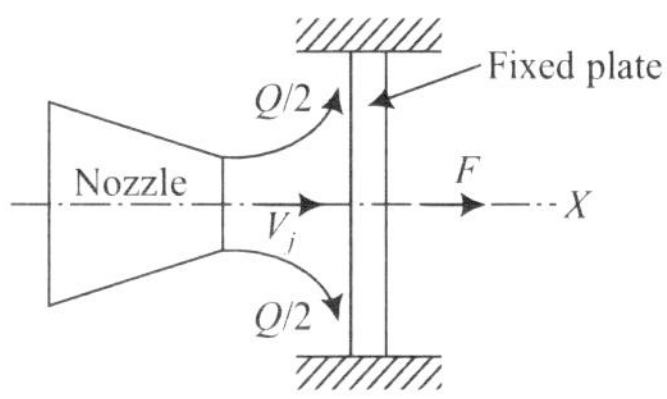

Fig. 2.10 Jet force on normal fixed plate

Example A jet of water 75 mm diameter strikes a normal fixed plate with a velocity of 200 m/s. Find the force exerted by water jet on the plate.

Solution $F_j = \rho \dot{Q} V_j = \rho A_j V_j V_j = \rho \frac{\pi}{4} D_j^2 V_j^2$

$$= 1000 \times \frac{\pi}{4}\left(\frac{75}{1000}\right)^2 \times (20)^2 = 1766.8 \text{ N}$$

$$= \mathbf{1.767 \text{ kN}}$$

2.11.2 Inclined Plate

A fluid jet with a velocity V_j strikes a fixed plate inclined at an angle θ with the direction of jet.

The components of jet velocity,

$$V_{jx'} = V_j \sin\theta$$

$$V_{jy'} = V_j \cos\theta = 0$$

The jet force,

$$F_j = -\rho \dot{Q} V_{jx'}$$

$$= -\rho \dot{Q} V_j \sin\theta.$$

The jet force has components in X-direction and Y-direction.

$$F_{jx} = F_j \sin\theta = -\rho Q V_j \sin^2\theta$$

$$F_{iy} = -\rho \dot{Q} V_j \sin\theta \cos\theta = 0$$

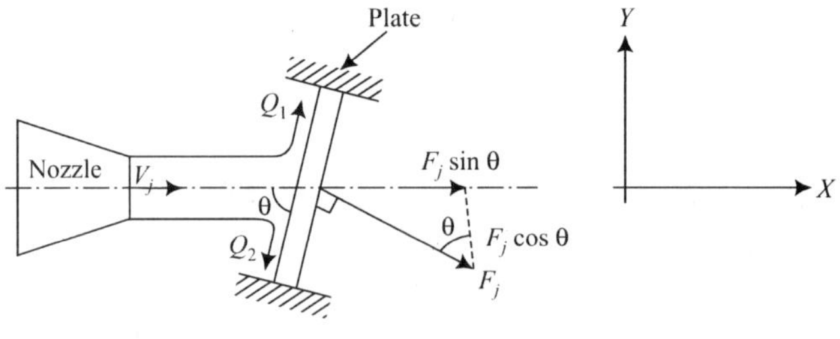

Fig. 2.11 Jet force on inclined fixed plate

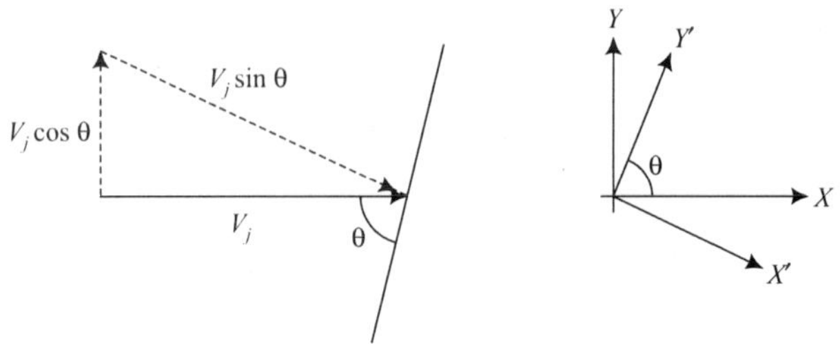

Fig. 2.12 Components of jet velocity

The volumetric discharge Q is divided into Q_1 and Q_2.

$$Q = Q_1 + Q_2$$

$$Q_1 = \frac{Q}{2}(1 + \cos\theta)$$

$$Q_2 = \frac{Q}{2}(1 - \cos\theta)$$

2.12 JET FORCE ON MOVING FLAT PLATE

The absolute velocity of jet issuing from the nozzle, $= V_j$

The velocity of moving plate in the direction of jet $= U$.

The jet force

$$F_j = \rho\dot{Q}(V_j - U)$$

The quantity of fluid mass striking the plate per second,

$$Q = A(V_j - U)$$

$$\therefore \quad F_j = \rho A(V_j - U)^2$$

The work done by the jet on the plate

$$W_j = F_j \cdot U = \rho\dot{Q}(V_j - U)U.$$

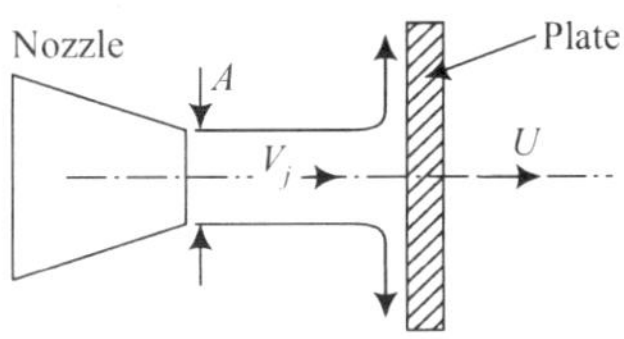

Fig. 2.13 Jet force on moving flat plate

The kinetic energy of jet

$$KE_j = \frac{1}{2}\dot{m}V_j^2 = \frac{1}{2}\rho\dot{Q}V_j^2 = \frac{1}{2}\rho A V_j^3$$

The efficiency of the system,

$$\eta = \frac{\text{Work done on the plate}}{\text{Kinetic energy of jet}} = \frac{W_j}{KE_j}$$

$$= \frac{\rho A(V_j - U)^2.U}{\frac{1}{2}\rho A V_j^3} = \frac{2}{V_j^3}[V_j^2 U + U^3 - 2U_j U^2]$$

For a given V_j, maximum efficiency.

$$\frac{d\eta}{dU} = \frac{2}{V_j^3}(V_j^2 + 3U^2 - 4V_j U) = 0$$

$$\frac{2}{V_j^3} \neq 0$$

$$\therefore \quad V_j^2 + 3U^2 - 4V_jU = 0$$

$$(V_j - U)(V_j - 3U) = 0$$

$$V_j = U \text{ gives } W_j = 0$$

$$\therefore \quad V_j = 3U$$

The maximum work done,

$$W_{j\max} = \rho A\left(V_j - \frac{V_j}{3}\right)^2 \times \frac{V_j}{3} = \frac{4}{27}\rho A V_j^3$$

$$\eta_{\max} = \frac{\frac{4}{27}\rho A V_j^3}{\frac{1}{2}\rho A V_j^3} = \frac{8}{27}$$

Special Case

When a series of plates is arranged, i.e., flat plates mounted on the periphery of a wheel, the effective distance between nozzle and plate remains constant. The quantity of fluid mass striking the plate per second,

$$\dot{m} = \rho A V_j$$

The jet force acting on the plate,

$$F_j = \rho A V_j (V_j - U)$$

$$W_j = F_j \cdot U$$

$$= \rho Q (V_j - U) U$$

The efficiency of the system,

$$\eta = \frac{W_j}{KE_j} = \frac{\rho Q(V_j - U)U}{\frac{1}{2}\rho Q V_j^2}$$

$$= \frac{2(U_j - U)U}{V_j^2} = \frac{2V_jU - 2U^2}{V_j^2}$$

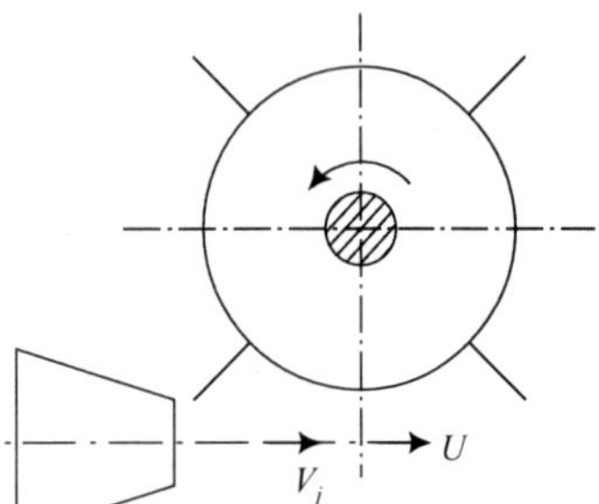

Fig. 2.14 Flat plates mounted on the periphery of a wheel

$$\frac{d\eta}{dU} = \frac{2V_j - 4U}{V_j^2} = 0$$

$$\therefore \quad V_j = 2U \quad \text{or} \quad U = \frac{V_j}{2}$$

$$\eta_{max} = \frac{2\left(V_j - \frac{V_j}{2}\right)\frac{V_j}{2}}{V_j^2} = 0.5 \quad \text{or} \quad 50\%.$$

2.13 JET FORCE ON CURVED PLATE WHEN JET STRIKES TANGENTIALLY

2.13.1 Stationary Vane

The fluid jet enters the curved vane with absolute velocity V_1 glides along the smooth inner surface and leaves with absolute velocity, V_2, making angles α_1 and α_2 with the horizontal direction.

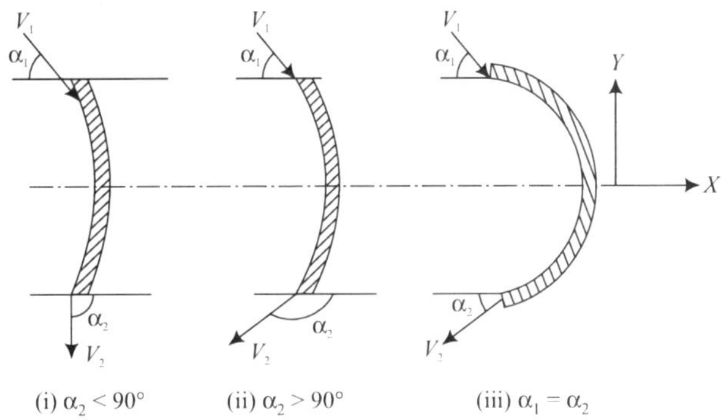

Fig. 2.15 Jet force on stationary vane

The velocity components in the X-direction are:

$$V_{X_1} = V_1 \cos \alpha_1$$

$$V_{X_2} = V_2 \cos \alpha_2.$$

The force exerted by the jet on the curved vane,

$$F_X = \rho Q(V_1 \cos \alpha_1 - V_2 \cos \alpha_2)$$

But $\qquad Q = AV_1$

where A = cross-section area of jet.

$$\therefore \qquad F_X = \rho A V_1(V_1 \cos \alpha_1 - V_2 \cos \alpha_2)$$

(i) If $\alpha_2 > 90°$

$$F_X = \rho A V_1(V_1 \cos \alpha_1 + V_2 \cos \alpha_2)$$

In order to get more power, $\alpha_2 > 90°$.

(ii) For symmetrical vane, $\alpha_1 = \alpha_2 = \alpha$ and $V_1 = V_2 = V$

$$F_X = \rho A V^2 (\cos\alpha + \cos\alpha) = 2\rho A V^2 \cos\alpha.$$

The hydraulic thrust in Y-direction

$$F_y = \rho A V^2 (\sin\alpha - \sin\alpha) = 0$$

(iii) For semicircular vane, $\alpha_1 = \alpha_2 = 0$.

$$\therefore \quad F_X = \rho A V^2 (\cos 0° + \cos 0°) = 2\rho A V^2$$

$$F_y = \rho A V^2 (\sin 0° - \sin 0°) = 0$$

Semicircular vanes give maximum hydraulic force and zero hydraulic thrust.

2.13.2 Moving Vane

A fluid jet of cross-sectional area A strikes a curved vane with an absolute velocity V_1. The curved vane moves with a velocity U in X-direction. The tip angles of vane are α_1 and α_2 at inlet and exit respectively. The velocity triangles at inlet and outlet of vane are drawn.

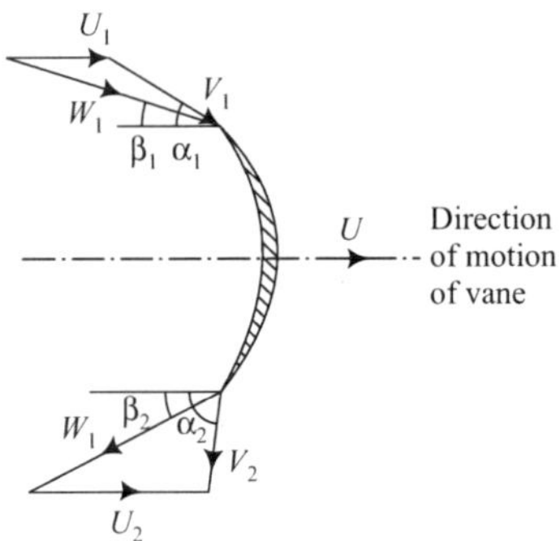

Fig. 2.16 Jet force on moving vane.

For frictionless vane, the relative velocity does not change,

$$W_1 = W_2$$

$$Q = A(V_1 - U)$$

$$F_X = \rho A(V_1 - U)(V_1 \cos\alpha_1 - V_2 \cos\alpha_2)$$

If $\alpha_2 > 90°$

$$F_X = \rho A(V_1 - U)(V_1 \cos\alpha_1 + V_2 \cos\alpha_2)$$

In actual hydraulic machines, a series of vanes is mounted on the periphery of a wheel.

2.14 JET PROPULSION OF SHIPS

2.14.1 Jet Reaction

A water jet of area A issues with a high velocity V. The jet velocity,

$$V = Kv\sqrt{2gh}$$

The fluid discharge,

$$Q = AV$$

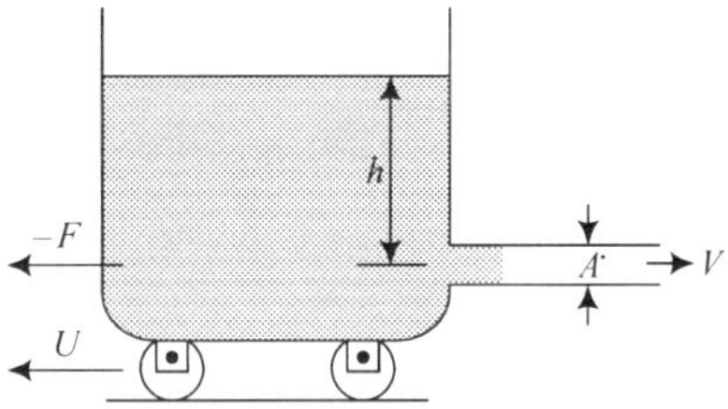

Fig. 2.17 Jet reaction

The force of reaction,

$$F = \rho Q(V_{X_2} - V_{X_1}) = \rho QV$$

The propulsive force exerted on the nozzle or tank = $-F$.

This jet reaction can be used to propel aircraft, rocket, ships or submarine. A nozzle is fitted to these machines.

2.14.2 Propulsion of Ships/Boats

A pump in the ship sucks water from the sea and discharges through a nozzle with a velocity V. The ship or boat moves in the opposite direction with a velocity U.

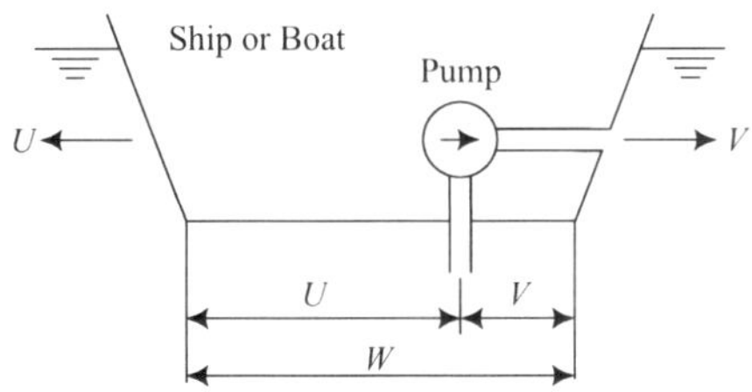

Fig. 2.18 Ship propulsion

The relative velocity of water jet with respect to motion of ship,

$$W = U + V$$

The jet force

$$F = \rho QV$$

Work done by the jet on the ship

$$W.D. = \rho QV.V = \rho Q(W - U)U$$

The energy input,

$$E = \frac{1}{2}\rho QW^2$$

The efficiency of the system,

$$\eta = \frac{W.D.}{E} = \frac{\rho Q(W-U).U}{\frac{1}{2}\rho Q W^2} = \frac{2(W-U)U}{W^2}$$

For maximum efficiency of the system

$$\frac{d\eta}{dU} = 0$$

$$\therefore \qquad \frac{d}{dU}\frac{2(WU-U^2)}{W^2} = 0$$

But $$\frac{W^2}{2} \neq 0$$

$$\therefore \qquad U = \frac{W}{2}$$

$$\eta_{\max} = \frac{2\left(W - \frac{W}{2}\right)\frac{W}{2}}{W^2} = \frac{1}{2} = 50\%$$

QUESTION BANK NO. 2

1. How are the hydraulic machines classified?
2. Derive Euler's equation applied to fluid machines.
3. Derive the continuity equation in Cartesian coordinates.
4. Discuss the following aspects of hydraulic flow:
   (a) Closed and open systems.
   (b) Concept of continuum.
   (c) Lagrangian and Eulerian methods.
   (d) Streamline and stream tube
5. Derive momentum and angular momentum equation.
6. What are the applications of momentum and angular momentum equations in fluid machines?
7. Establish the Bernoulli's equation from Euler's equation of motion.
8. Explain water jet theory and how are the following principles established.
   (a) Impulse principle
   (b) Reaction principle
   (c) Propulsion principle
9. Derive equation of a jet force for moving flat plate and moving curved vane.

TUTORIAL SHEET NO. 2

1. A 25 cm diameter pipe carries lubrication oil of 0.9 specific gravity at a velocity of 3 m/s. Find the mass flow rate of oil. What will be the oil velocity at another section where diameter is reduced to 20 cm.

Solution Section 1:

$$D_1 = 25 \text{ cm} = 0.25 \text{ m}$$

$$A_1 = \frac{\pi}{4} D_1^2 = \frac{\pi}{4}(0.25)^2$$

$$= 0.049 \text{ m}^2.$$

$$V_1 = 3 \text{ m/s}.$$

$$\rho = 0.9 \times 1000 = 900 \text{ kg/m}^3.$$

$$\dot{m} = \rho A_1 V_1 = 900 \times 0.049 \times 3$$

$$= \mathbf{132.23\ kg/s} \quad \text{Ans.}$$

Section 2

$$D_2 = 20 \text{ cm} = 0.20 \text{ m}$$

$$A_2 = \frac{\pi}{4}(0.20)^2 = 0.314 \text{ m}^2$$

Applying equation of continuity at sections 1 and 2,

$$A_1 V_1 = A_2 V_2$$

$$\therefore \quad V_2 = \frac{A_1 V_1}{A_2} = \frac{0.049 \times 3}{0.0314} = \mathbf{4.68\ m/s} \quad \text{Ans.}$$

2. Water under a pressure of 29.43 N/cm^2 (g) and velocity of 2 m/s is flowing through a 5 cm diameter pipe. Find the total head of water at a section 5 m above datum line.

Solution Pressure head $= \dfrac{p}{\rho g} = \dfrac{29.43 \times 10^4}{1000 \times 9.81} = 30$ m.

$$\text{Kinetic head} = \frac{V^2}{2g} = \frac{2 \times 2}{2 \times 9.81} = 0.204 \text{ m}$$

$$\text{Datum head} = Z = 5 \text{ m}.$$

$$\therefore \quad \text{Total head} = 30 + 0.204 + 5 = \mathbf{35.204\ m} \quad \text{Ans}$$

3. A jet of water 60 mm in diameter, having a velocity of 20 m/s, strikes a flat plate inclined at an angle of 30° to the axis of the jet. The plate moves at 5 m/s in the direction of the jet.

Calculate the normal force exerted on the plate, work done per second and efficiency of the system.

Sketch the arrangement showing velocity and force components.

Solution Refer Figs. 2.11 and 2.12.

The normal force on the moving plate,

$$F_{j_n} = \rho A(V_j - V_b)^2 \sin\alpha$$

$$= 1000 \times \frac{\pi}{4}(0.06)^2 \times (20-5)^2 \sin 30°$$

$$= \mathbf{318\ N} \quad \text{Ans.}$$

The force component in the direction of jet

$$F_{j_x} = F_{j_n} \sin\alpha = 318 \times 0.5 = \mathbf{159\ N} \quad \text{Ans.}$$

Work done per second,

$$W = F_{j_x} V_b = 159 \times 5 = 795 \text{ Nm/s}$$

$$= \mathbf{0.795\ kW} \quad \text{Ans.}$$

Kinetic energy supplied by the jet

$$KE_j = \frac{1}{2} m V_j^2 = \frac{1}{2}(\rho A V_j) V_j^2$$

$$= \frac{1}{2}\rho A V_j^3 = \frac{1}{2} \times 1000 \times (0.06)^2 \times 20^3$$

$$= 11309.7 \text{ Nm/s.}$$

Efficiency of the system,

$$\eta = \frac{W}{KE_j} = \frac{795}{11309.7} \times 100 = \mathbf{7\%} \quad \text{Ans.}$$

4. Show that angle of swing of a vertical hinged plate is given by.

$$\sin\theta = \frac{\rho A V_j^2}{W}$$

where V_j = velocity of jet striking the plate
A = cross-sectional area of jet
W = weight of plate
θ = angle of swing of plate from vertical

Solution A plate *AB* is hinged at *A* and is free to swing. Initially, the plate is vertical and a water jet strikes it normally at mid-point *G* with a velocity,

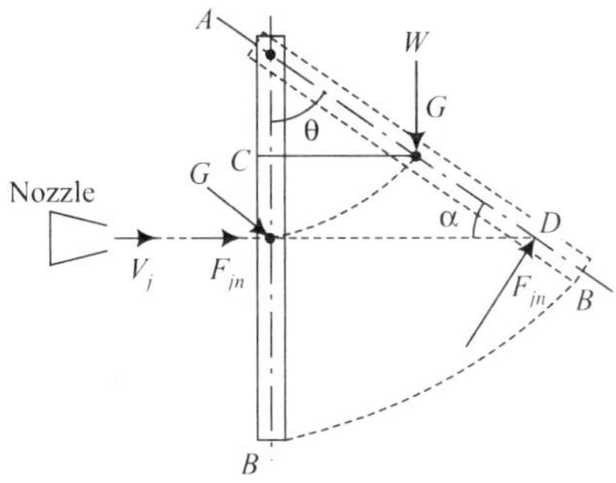

V_j. The plate swings through an angle θ. The plate experiences the following forces

1. Weight W of plate acting vertically through G.
2. Normal water jet force, F_{j_n},

$$F_{j_n} = \rho A V_j^2 \sin \alpha$$

where α = angle of jet with the centre line of plate.

$$\therefore \quad F_{j_n} = \rho A V_j^2 \sin (90° - \theta)$$

$$= \rho A V_j^2 \cos \theta.$$

The plate is in equilibrium under the two forces. Taking moments about point A.

$$F_{j_n}(AD) = W(CG)$$

$$AD = \frac{AG}{\cos\theta}$$

$$CG = AG \sin \theta.$$

$$\therefore \quad F_{j_n} \frac{(AG)}{\cos\theta} - W(AG) \sin\theta$$

$$\rho A V_j^2 \cos\theta \frac{(AG)}{\cos\theta} = W(AG) \sin\theta.$$

$$\therefore \quad \sin\theta = \frac{\rho A V_j^2}{W}$$ **Proved.**

PART II

TURBINES

CHAPTER 3

Pelton Turbine

3.1 INTRODUCTION

Hydroelectric power plants utilize the potential energy of water to drive a turbine which, in turn, runs the generator to produce electricity. The hydro turbines are designed based on water head and discharge available at site. For high head and low discharge (water flow), Pelton turbines are used which run on impulse principle. For low head and large discharge, Kaplan and Propeller turbines are used which work on reaction principle. For medium head and discharge, Francis turbine is used which also works on reaction principle. The Pelton turbine is usually designed as horizontal shaft whereas Kaplan and Francis turbines are designed as vertical shaft machines. A special reversible turbine pump has been designed by a Swiss engineer and is named after him as Deriaz turbine. The comparative study of various types of hydro turbines is discussed in the following paragraph.

3.2 COMPARATIVE STUDY OF HYDRO TURBINES

3.2.1 Specifications

The hydraulic turbines have been classified and main specifications are given in Table 3.1.

Table 3.1 Classification of hydraulic turbines

S.No.	*Parameters*	*Pelton Turbine*	*Francis Turbine*	*Kaplan Turbine*
1.	Originator	Lester Allen Pelton, USA	James B. Francis	Dr. Victor Kaplan
2.	Principle of operation	Impulse	Reaction	Reaction
3.	Head and discharge	High head, low discharge	Medium head Medium discharge	Low head, large discharge
4.	Maximum head (m)	300-2000	30-500	2-70
5.	Maximum power (MW)	250	720	225
6.	Maximum wheel diameter (m)	5.5	10	10
7.	Flow direction	Tangential	Mixed Flow (Radial Inward)	Axial
8.	Specific speed	4-70	60-400	300-1100
9.	Shaft axis	Horizontal	vertical	vertical

The layout drawings of various turbines are shown in Fig. 3.1.

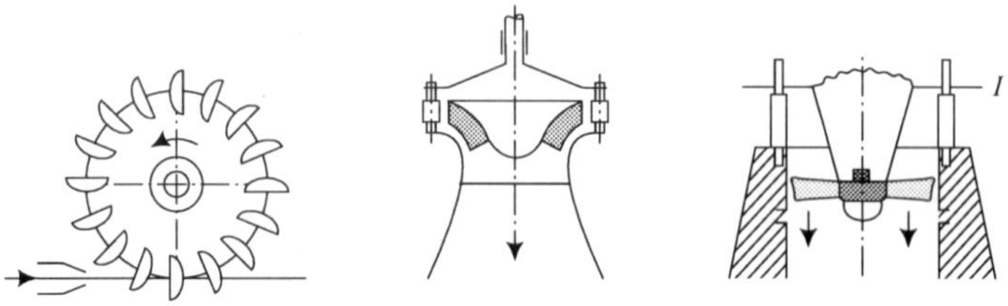

(a) Pelton turbine (b) Francis turbine (c) Kaplan turbine

Fig. 3.1 Layout of hydraulic turbines

3.2.2 Performance Curves

The efficiency curves of various turbines are plotted against normalized water discharge in Fig. 3.2. Francis turbine is suitable for constant load operation whereas Kaplan turbine is suitable for variable load operation. Pelton turbines have lower efficiency but can be used as variable load turbine.

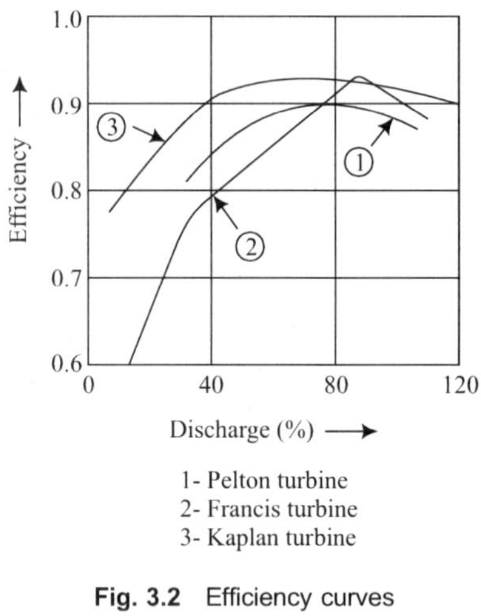

Fig. 3.2 Efficiency curves

3.2.3 Selection of Hydro Turbines

The following criteria are used for the selection of hydro turbine for a site to attain the maximum possible efficiency and economy.

1. Operating head

The present practice is to use Kaplan and Propeller type of turbines for heads upto 50 m. Francis turbines are selected for water heads from 50 m to 400 m. For heads greater than 400 m, Pelton turbine is used. The range of heads as mentioned is not rigid and may change if other conditions dominate to achieve economy.

2. Size of turbine and number of units

A large size of hydro turbine results in less number of units for a power plant. This leads to lower capital cost due to economy of size and lesser number of runners, penstocks, generators, etc. However, less number of hydrosets are not able to meet the fluctuating load requirements economically as the turbines may not be operating at optimum efficiency at part loads. In no case there should be less than two turbines to ensure reserve availability of capacity during plant shut down under normal and emergency conditions.

There is limitation on the size of type of turbine as given in Table 3.1.

3. Turbine Speed

High speed turbines are more economical as the size of turbine, generator, power house, etc. are smaller. The turbine speed should correspond to synchronous speed of generator, i.e.

$$N = \frac{120f}{p}$$

where f = grid frequency (50 cps)

p = number of poles of generator

The specific speed range for different turbines is given in Table 3.1. The specific speed,

$$N_S = \frac{N\sqrt{P}}{H^{5/4}}$$

where N = turbine operating speed, (rpm)

P = turbine output (kW)

N = water head (m)

4. Load characteristics

The performance characteristics of various turbines are shown in Fig. 3.2. Kaplan and Pelton turbines have flat efficiency curves and can operate at high efficiency for large range of part loads. Francis turbine is not suitable to operate at part loads as efficiency falls appreciably. However, Francis turbines have high efficiency when operating as base load turbine.

5. Height of Installation

The cost of excavation can be saved, if the turbine can be installed at higher level, Pelton turbines are generally horizontal shaft turbines requiring little excavation. Francis and Kaplan turbines are generally installed vertically and are fitted with draft tube to avoid cavitation. This requires excavation for installation of draft tubes. For these machines also height of installation should be more to save excavation cost.

3.3 PELTON TURBINE

Pelton wheel is an impulse turbine used for high head installation. It has evolved from Himalayan Mill which is still widely used in the high mountains from Afghanistan to Burma for grinding of grains. The pressure energy of water available in high reservoirs is converted into velocity head. The water is carried from reservoirs through penstocks. A nozzle is fitted at the end of penstock which issues a free and compact jet of water. The water jet sticks a series of buckets mounted around a circular wheel. The buckets are designed in the form of two-lobe ellipsoidal discs for high conversion efficiency and balanced hydraulic thrust. Each bucket has a ridge or splitter in the middle to divide the jet into two equal streams. The symmetry of the bucket ensures no axial face on the shaft bearings. The wheel rotates and supplies torque or mechanical power to the shaft carrying the wheel.

The Pelton turbine is a very reliable and efficient in operation at high heads. At low heads, the rate of water flow has to be increased for a given power output. The high flow rate leads to big jet diameter and high jet velocity resulting in:

1. Large runner diameter
2. Low peripheral velocity of runner

The increase of runner diameter and decrease of its velocity requires a bulky, slow speed machine. For low heads and large flow rate, reaction turbines are more suitable.

3.3.1 Main Components

The main components of a Pelton turbine are:

1. Water inlet system
2. Runner assembly
3. Casing
4. Brakes

1. Water Inlet system

(a) Penstock A large pipe or conduit made of steel, concrete or wood carries water from high level reservoir to the powerhouse. It is provided with suitable anchor blocks to take the hydraulic thrust. Screens or trash racks are provided at the inlet to prevent entry of debris. There is a stop valve and a control valve provided in the penstock to regulate the flow of water. The penstock ends in a nozzle.

(b) Spear and Nozzle At the end of the penstock, there are one or several nozzles which convert the pressure energy into high speed jet. A spear or needle moves to and fro in the nozzle to vary the outlet annular area, thereby controlling the water flow rate depending upon the load on the turbine-generator. The

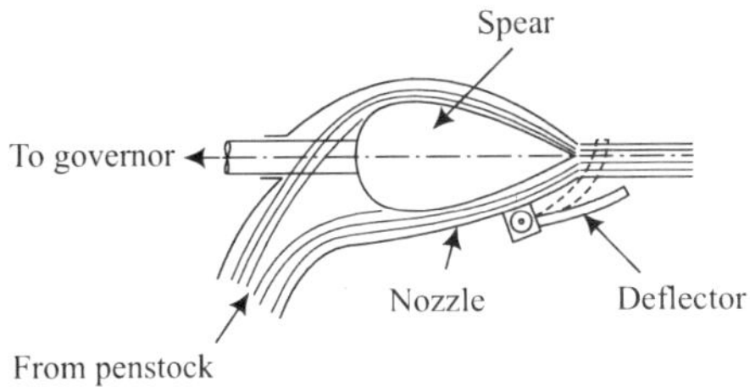

Fig. 3.3 Spear and nozzle.

movement of the spear is automatically controlled by hydraulic system of governing system.

The runner speed has to be maintained constant at all loads and variable heads.

(i) When the load on the generator decreases or head increases, the turbine speed will increase. The speed governor operates to push the spear forward into the nozzle reducing the annular area and water discharge. The turbine speed is reduced to synchronous speed.

(ii) When the load on the generator increases or head decreases, the turbine speed falls. The speed governor operates to withdraw the spear backward. The annular area of nozzle increases and also increases the water discharge. The turbine speed is increased to synchronous speed.

(iii) *Protection System* In case of sudden load throw off, quick closing of nozzle can cause bursting of inlet pipe. This is avoided by the use of one of the following devices which operate along with slow closing of main nozzle.

(1) By-pass nozzle to discharge extra water without striking the runner buckets.

(2) The governor swings a deflector plate in between the nozzle and runner buckets when load is thrown off.

(3) *Seewer Jet Diffuser* When the load is thrown off, a number of blades move forward towards the spear needle and whirl the flow and thus dissipate the water energy.

2. Runner assembly

The runner assembly consists of a circular wheel which is shrunk fit and keyed to a shaft supported on bearing. An odd number of specially shaped buckets (not less than 15) are bolted to the wheel periphery at equal pitch. Sometimes the wheel and buckets are cast together.

The buckets are double hemispherical cup-shaped castings of cast iron, bronze or stainless steel or cast steel. The inner surface of buckets is finely polished to reduce flow friction of water. The jet strikes the splitter ridge at the centre and flow is equally deviated on both sides, thus eliminating the end thrust.

The water jet after impinging on the buckets is deflected through an angle of about 165° (instead of 180°) to avoid its splashing on the incoming bucket and retarding the runner motion. A recess or notch is provided at the bottom of the bucket not to interfere with the water jet till it is in proper position.

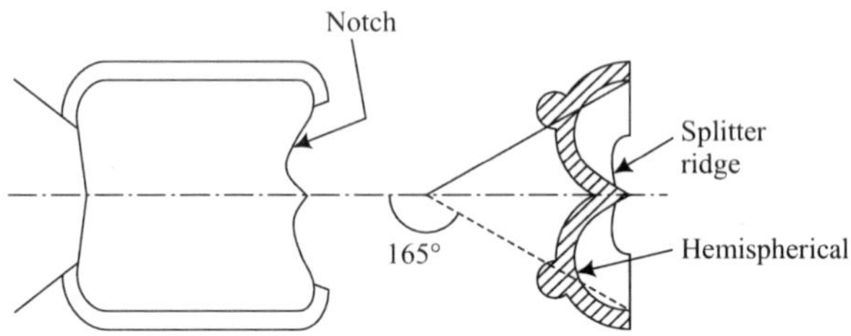

Fig. 3.4 Bucket

3. Casing

The casing has atmospheric pressure and has no hydraulic function. The water outflow from bucket issues in the form of a strong splash and scatters in all direction. The casing avoids the water splash and guides the water to the tail race. It also acts as safe gaurd against accidents.

The casing is made in two halves with a seal at the joint. Baffles are also provided to reduce windage losses. The casing can be a cast or fabricated from steel plate.

4. Brakes

When the turbine stops, the runner can still continue revolving for quite sometime due to inertia. A hydraulic brake is provided to forcefully stop the runner in minimum time. The hydraulic brake consists of a nozzle of diameter 60% of main nozzle and is situated and positioned suitably, so that the brake jet strikes the bucket at its back.

3.4 LAYOUT ARRANGEMENTS

The sectional view of a single jet horizontal Pelton turbine is shown in Fig. 3.5. All the main components have been arranged. The following design changes are discussed below.

1. Number of jets

One nozzle Pelton turbines are widely used. However, large turbines may have more than one nozzle depending upon the specific speed to control the running speed. Multi-jets are essential when $N_S > 35$. The arrangement layout of a 4-jet Pelton turbine is shown in Fig. 3.6.

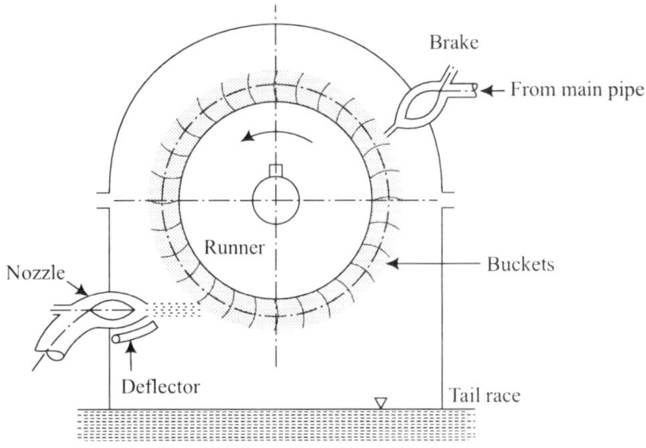

Fig. 3.5 Layout of Pelton turbine

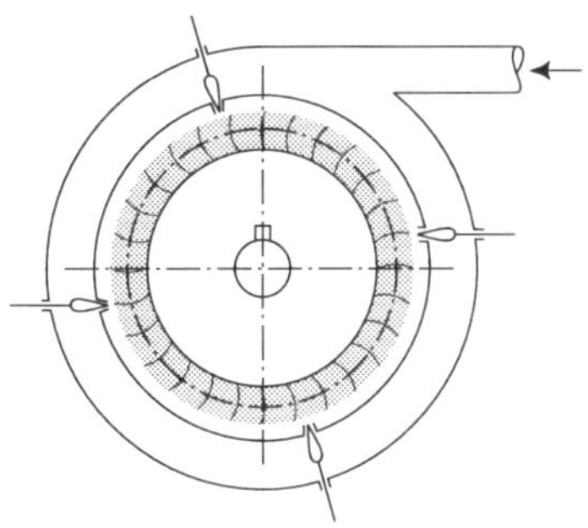

Fig. 3.6 Multi-jet Pelton turbine

2. Number of runners

The most common Pelton turbine layout is single overhung runner. However, double runner gives more power and higher speed. The various arrangements are shown below.

3. Shaft arrangement

Horizontal Pelton turbines are most common. However, horizontal shaft arrangement is used for Pelton turbines with one or two nozzle. For four or six nozzle turbines, vertical shaft arrangement is used.

3.5 VELOCITY DIAGRAM OF BUCKET

The inlet and outlet velocity triangles of the bucket are drawn to estimate the work output and efficiency of a Pelton wheel.

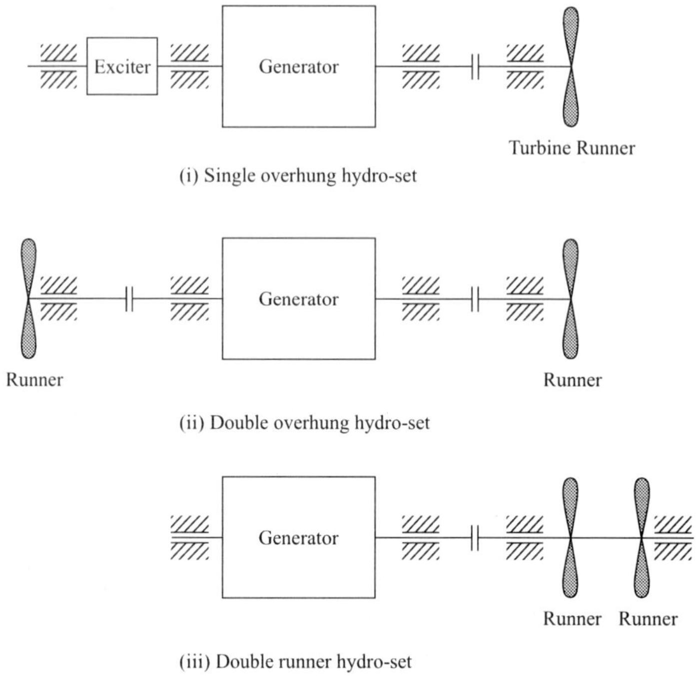

Fig. 3.7 Runner arrangement

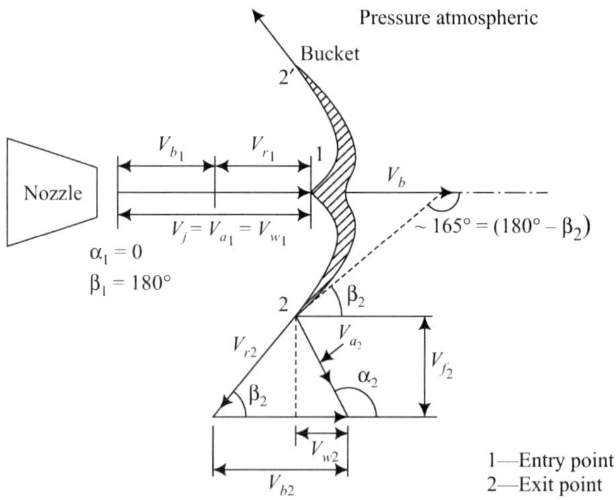

Fig. 3.8 Velocity diagrams

Blade is divided into two parts to avoid side thrust or axial thrust

V_j = Jet velocity [m/s]

V_a = Absolute velocity of water

V_b = Bucket or blade tip velocity

V_w = Whirl velocity

V_f = Flow velocity

V_r = Relative velocity

1. Inlet velocity triangle

Pelton wheel is an axial flow impulse turbine and the angle of entry of the jet is nearby zero.

$$\alpha_1 = 0 \quad \text{and} \quad \beta_1 = 180°$$

where α = Angle of absolute velocity with blade velocity

β = Angle of relative velocity with blade velocity

$$\therefore \qquad V_{a_1} = V_j = V_{w1}$$

$$V_{r_1} = (V_j - V_{b1})$$

$$V_b = V_{b_1} = V_{b_2} = \frac{\pi DN}{60}$$

2. Outlet velocity triangle

The outlet velocity triangle depends upon the speed of the runner. The values of α_2 and V_{w_2} will change as shown.

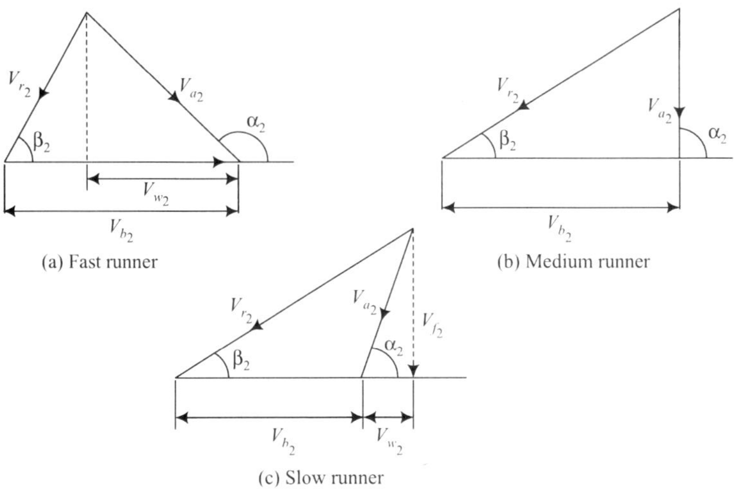

Fig. 3.9 Outlet velocity triangles

Type of Runner	*Angle* α_2	*Velocity* V_{w_2}
Slow Runner	$< 90°$	$-$ve
Medium Runner	$= 90°$	0
Fast Runner	$> 90°$	$+$ve

The buckets are polished to fine finish from inside to reduce friction.

$$V_{r_2} = KV_{r_2}, \quad \text{where} \quad K \approx 0.85$$

Under ideal conditions, assuming no friction

$$K = 1.0$$

$$V_{r_2} = V_{r_1}$$

$$V_{b_1} = V_{b_2} = V_b \quad (\because r_1 = r_2)$$

3. Work output

The kinetic energy of water at inlet,

$$KE_1 = \frac{Va_1^2}{2} \quad \text{[N-m/kg]}$$

The kinetic energy of water at outlet,

$$KE_2 = \frac{Va_2^2}{2}$$

$$\text{Energy transferred to rotor} = \left(\frac{Va_1^2}{2} - \frac{Va_2^2}{2}\right)$$

Power input to wheel or work done on the wheel,

$$W = (\overline{V}_{w_1} r_1 - \overline{V}_{w_2} r_2)\, w \qquad (\because W = Tw)$$

$$= (\overline{V}_{w_1} - \overline{V}_{w_2})\, wr \qquad (V_b = wr)$$

$$= (\overline{V}_{w_1} - \overline{V}_{w_2}) V_b$$

For fast runner

$$W = (V_{w_1} - V_{w_1}) V_b$$

For medium runner

$$W = 0$$

For slow runner

$$W = (V_{w_1} + V_{w_2}) V_b$$

4. Hydraulic efficiency

The hydraulic efficiency or blade efficiency

$$\eta_{\text{hyd}} = \frac{\text{Work done on wheel}}{\text{Jet power}}$$

$$= \frac{(V_{w_1} - V_{w_2})V_b}{\frac{V_j^2}{2}}$$

From velocity diagrams,

$$V_{w_1} = V_j$$

$$V_{w_2} = V_{b_2} - V_{r_2} \cos \beta_2$$

$$= V_{b_2} - KV_{r_1} \cos \beta_2 = V_{b_2} - K(V_j - V_{b_1}) \cos \beta_2$$

$$\therefore \quad \eta_{\text{hyd}} = \frac{[V_j + K(V_j - V_{b_1}) \cos \beta_2 - V_b]V_b}{V_j^2/2} \tag{1}$$

For a given runner, β_2, V_j and K are constants.

For estimation of maximum hydraulic efficiency,

$$\frac{d(\eta_{\text{hyd}})}{dV_b} = 0$$

From equation (1)

$$\eta_{\text{hyd}} = \frac{2V_b}{V_j} + \frac{2KV_b}{V_j} \cos \beta_2 - \frac{2KV_b^2 \cos \beta_2}{V_j^2} - \frac{2V_b^2}{V_j^2}$$

$$\therefore \quad \frac{d(\eta_{\text{hyd}})}{dV_b} = \frac{2}{V_j} + \frac{2K \cos \beta_2}{V_j} - \frac{4KV_b \cos \beta_2}{V_j^2} - \frac{4V_b}{V_{j^2}} = 0$$

$$2 + 2K \cos \beta_2 - 4K \frac{V_b}{v_j} \cos \beta_2 - \frac{4V_b}{V_j} = 0$$

$$2 + 2K \cos \beta_2 = \frac{2V_b}{V_j} (2 + 2K \cos \beta_2)$$

$$\frac{V_b}{V_j} = \frac{1}{2}$$

$\therefore$ Speed ratio, $$\frac{V_b}{V_j} = \frac{V_b}{\sqrt{2gHa}} = 0.5$$

Actual speed ratio varies between 0.46 and 0.47

$$\therefore \quad V_b = 0.5\sqrt{2gHa} = \frac{\pi D_m N}{60}$$

$$D_m = \frac{60 V_b}{\pi N}$$

The blade efficiency curve is shown in Fig. 3.10.

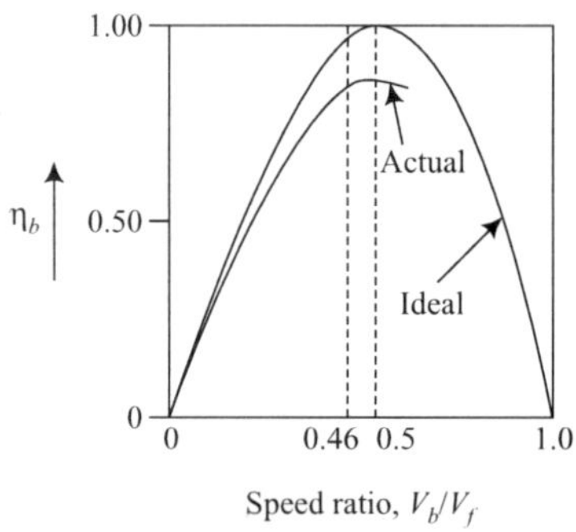

Fig. 3.10 Blade efficiency curve

5. Jet ratio

Jet ratio is an important characteristic of Pelton turbine and is defined as

$$m = \frac{D_m}{d_j}$$

where D_m = Mean diameter of runner

d_j = jet diameter

m varies from 6.5 to 20. For maximum η_{hyd}, m should be taken between 10 and 14.

3.6 TURBINE LOSSES AND EFFICIENCY

For performance analysis, the various losses in the turbine are estimated and plotted in a Sankey diagram. The energy balance sheet is also prepared.

There are three types of losses.

1. Volumetric losses

Volumetric losses arise as full volume of the jet does not strike the bucket. The volumetric losses may be of the order of 1% to 3%.

2. Hydraulic losses

These losses are due to the following factors:

(i) Exit losses due to kinetic energy of water leaving the bucket $\left(\frac{Va_2^2}{2}\right)$.

(ii) Friction and shock losses on the bucket due to improper inlet blade angle and surface finish.

3. Mechanical losses

The mechanical losses include:

(i) Bearing friction

(ii) Windage losses

The mechanical losses may be 1% to 3%.

3.6.1 Sankey Diagram

The Sankey diagram for a Pelton turbine is shown in Fig. 3.11.

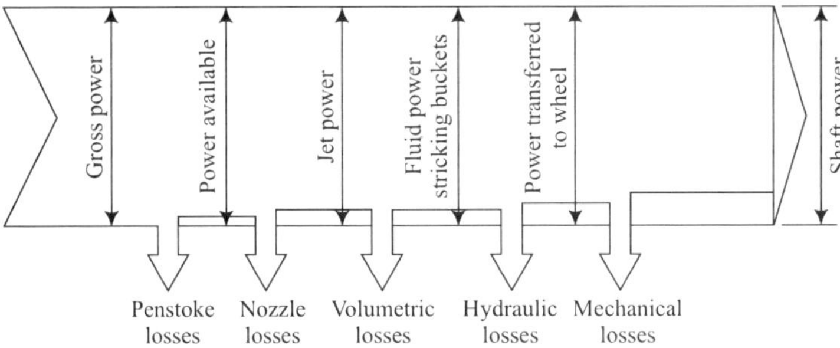

Fig. 3.11 Sankey diagram

$$\eta_{nozzle} = \frac{\text{Jet power}}{\text{Available power}} = \frac{\rho Qg \frac{V_j^2}{2g}}{\rho Qg\, Ha} \quad (\approx 97\%)$$

$$\eta_{vol} = \frac{\text{Kinetic energy of fluid striking the bucket}}{\text{Jet power}}$$

$$= \frac{Q - \Delta Q}{Q} \quad (97 - 99\%)$$

$$\eta_{hyd} = \frac{\text{Power transferred to wheel}}{\text{Kinetic energy of fluid striking the bucket}}$$

$$\eta_{\text{hyd}} = \frac{\rho(Q-\Delta Q)\left(\frac{Va_1^2}{2} - \frac{Va_2^2}{2}\right)}{\rho(Q-\Delta Q)V_j^2/2} = \frac{(\overline{V}_{w_1} - \overline{V}_{w_2})V_b}{V_j^2/2}$$

V_{w_1} and V_{w_2} are measured from velocity diagrams.

$$\eta_{\text{hyd}} = \frac{(V_{a_1}^2 - V_{a_2}^2)}{V_j^2} = 1 - \left(\frac{Va_2}{V_j}\right)^2$$

$$\eta_{\text{mech}} = \frac{\text{Shaft power output}}{\text{Power transferred to wheel}} \quad (97 \text{ to } 99\%)$$

$$\eta_{\text{overall}} = \eta_{\text{turbine}} = \eta_{\text{nozzle}} \times \eta_{\text{vol}} \times \eta_{\text{hyd}} \times \eta_{\text{mech}} \quad (90 \text{ to } 92\%)$$

3.6.2 Energy Balance Sheet

A typical energy balance diagram is plotted in Table 3.2.

Table 3.2 Energy balance (Basis MW)

Credit		*Debit*	
Description	*MW*	*Description*	*MW*
1. Gross Power $P_g = \frac{\rho Qg Hg}{10^6}$ 2. Power Available $P_a = (P_g - P_{\Delta h_f})$ $= \frac{\rho Qg Ha}{10^6}$		1. Transmission losses $P_{\Delta h_f} = \frac{\rho Qg \Delta H_f}{10^6}$ 2. Brake Power $P_s = \frac{2\pi NT}{60 \times 10^6}$ 3. Nozzle losses $P_{\Delta h_n} = (P_a - P_j)$ 4. Volumetric losses $P_{\Delta h_{val}} = (1 - \eta_{vol})P_j$ 5. Hydraulic losses $P_{\Delta h_{hyd.}} = (1 - \eta_{hyd})P_j \eta_{vol}$ 6. Mechanical losses $P_{\Delta h_{mech}} = (1 - \eta_m)P_j \eta_{vol} \eta_{hyd}$	
Total		*Total*	

3.7 DESIGN OF PELTON TURBINE

The hydraulic design of a Pelton turbine is based on a large number of experimental and imperical relations and coefficients determined from dimensional analysis and good engineering practice.

1. Turbine power

The turbine power

$$p_t = P_a \eta_{\text{overall}}$$

Where P_a = Available Power (W)

$= \rho Q g Ha.$

Q = Volumetric flow rate (m^3/s)

Ha = Available head (m)

η_{overall} = Overall efficiency of turbine (85 to 90%)

2. Nozzle and jet size

The angle of nozzle = 2β

Angle of spear = 2α

α is 22.5 to 25°

β is 28.5 to 42°

The jet flow,

$$Q = \frac{\pi}{4} d_1^2 V_1 \quad (\text{m}^3/\text{s})$$

$$V_1 = K_{v_1}\sqrt{2gH}$$

where H = Head (m)

K_{v_1} = coefficient of velocity (0.98 – 0.99) at vena contracta of jet.

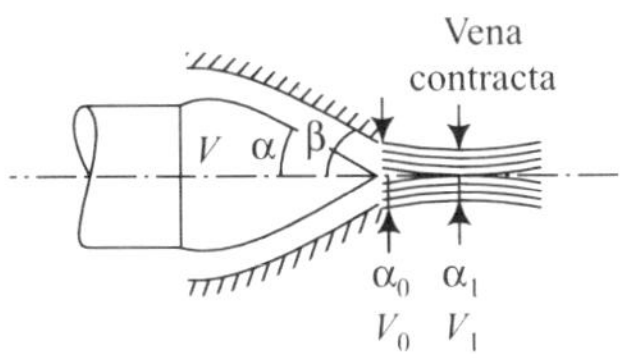

Fig. 3.12 Nozzle and jet size

Also $$Q = \frac{\pi}{4} d_0^2 V_0$$

∴ $$V_0 = K_{v_0}\sqrt{2gH}$$

K_{v_0} = coefficient of velocity at nozzle exit (0.81 – 0.83)

The area of nozzle exit,

$$A_0 = \frac{\pi}{4} d_0^2$$

Let $A_0 = AS + BS^2$

where A and B are constants and S is maximum spear travel

where
$$A = 2\pi r_0 \sin\alpha \frac{(\beta - \alpha)}{\sin(\beta - \alpha)}$$

and
$$B = \frac{\pi(\beta - \alpha)(\sin\beta - \sin\alpha)\sin^2\alpha}{\sin^2(\beta - \alpha)}$$

3. Number of jets

A Pelton wheel ordinarily has one nozzle and jet. However, more jets are employed when more power is produced by the runner.

The number of jets is decided by specific speed of the runner which is defined as follows.

The specific speed,

$$N_S = \frac{N\sqrt{P_t}}{H^{5/4}}$$

where N = Rotor speed (rpm)

P_t = Turbine power

H = Available head

The maximum value for a single jet Pelton wheel is limited to 35. If N_S is more, then the number of jets are increased.

There is another dimensionless quantity (K_S) used to decide number of jets.

$$K_S = \frac{QN^2}{V_j^3}$$

where Q = Volumetric flow rate (m^3/s)

V_j = Jet velocity

$V_j = \sqrt{2gH}$

$$K_S = \left(\frac{N_s}{32.2}\right)^2 \text{ for turbine efficiency, } \eta_t = 90\%$$

For single jet Pelton turbine, $K_S = 1$.

For Pelton turbine with one wheel and 2 jets, total power will be $2P_t$.

$$\therefore \qquad N_{S_2} = \frac{N\sqrt{2P_t}}{H^{5/4}} = \sqrt{2}\ N_{S_1}$$

Observations

1. For single-jet turbine, $K_S < 1$
2. For two-jet turbine, K_S lies between 1 and 2
3. For three-jet turbine, K_S lies between 2 and 3
4. Maximum number of jets used so far;

   2 jets for horizontal shaft turbine

   6 jets for vertical shaft turbine
5. Governing of turbine with multi-jets becomes complex

4. Runner diameter

The runner diameter of the turbine is also called pitch circle diameter measured from the centre of the buckets.

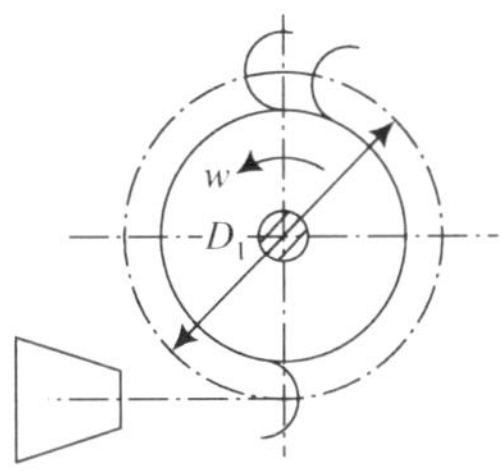

Fig. 3.13 Pitch diameter

The circumferential velocity or peripheral velocity,

$$U_1 = \phi\sqrt{2gH}$$

where ϕ = speed ratio

$$= \frac{U_1}{\sqrt{2gH}}$$

The speed ratio varies between 0.44 to 0.46 with average value of 0.45.

$$U_1 = \frac{\pi D_1 N}{60} \quad \text{(m/s)}$$

Now
$$\frac{\pi D_1 N}{60} = \phi\sqrt{2gH}$$

$$\therefore \qquad D_1 = \frac{60}{\pi} \times 0.45\sqrt{2g}\,\frac{\sqrt{H}}{N} = \frac{38.06\sqrt{H}}{N}$$

The rotor speed is decided by the synchronous speed of the generator

$$N = \frac{60f}{p}$$

where f = grid frequency

= 50 cps for India

p = no. of pair of poles of hydro-generator

The usual values of N are 500, 300 or 250 rpm. Higher values of N result in lower cost of turbine-generator set.

5. Rotor speed

The selection of rotor speed requires detailed techno-economical analysis.

The turbine efficiency is a function of specific speed. Lower value of N_S leads to higher turbine efficiency. N_S also depends upon rotor speed.

$$\eta_t = f(N_S)$$

$$N_S = \phi(N)$$

Higher rotor speed leads to:

1. Higher value of specific speed, N_S. This results in :
   (i) Compact and less expensive turbine.
   (ii) Jet diameter decreases, jet ratio increases and turbine efficiency increases.
   (iii) Multijet turbines. It is more expensive and governing is complicated.
2. Hydro-generator speed increases
   (i) Less number of poles are required.
   (ii) The cost of hydro-generator decreases.
3. Higher stresses in turbine and generator rotors
   (i) Superior materials are needed to match higher stresses
   (ii) Cost of turbine-generator increases.

6. Jet ratio

Jet ratio is defined as the ratio of the pitch circle diameter of runner and jet diameter.

Jet ratio, $$m = \frac{D_1}{d_1}$$

$$N_S = \frac{N\sqrt{P_t}}{H^{5/4}}$$

where $$N = \frac{60\phi\sqrt{2gH}}{\pi D_1}$$

$$P_t = \rho Q g H \eta_t.$$

$$Q = \frac{\pi}{4} d_1^2 \psi \sqrt{2gH}$$

$$\therefore \qquad P_t = \rho \frac{\pi}{4} d_1^2 \psi \sqrt{2gH} . gH \eta_t$$

$$N_S = \frac{60\phi\sqrt{2gH}}{\pi D_1} . \frac{\sqrt{\rho \frac{\pi}{4} d_1^2 \psi \sqrt{2gH} . gH \eta_t}}{H^{5/4}}$$

$$= \frac{260\sqrt{\eta_t}}{m}$$

The jet ratio is between 10 and 30. For maximum turbine efficiency $m = 12$ is adopted.

A small value of m results in a very few buckets and too close buckets spacing.

A large value of m results in bulky installation.

7. Bucket dimensions

The working dimensions of a bucket are specified in terms of jet diameter d_1.

$$L = (2.3 \text{ to } 2.8)d_1$$

$$B = (2.8 \text{ to } 3.2)d_1$$

$$T = (0.6 \text{ to } 0.9)d_1$$

Inlet angle,

$$\beta_1 \approx 5° \text{ to } 8°$$

Outlet angle,

$$\beta_2 = 10° \text{ to } 20° \text{ at the centre}$$

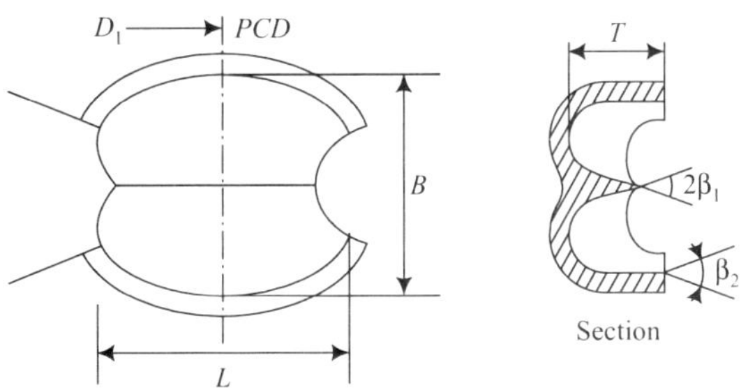

Fig. 3.14 Bucket dimensions

8. Number of buckets (z)

Minimum number of buckets on the periphery of a Pelton wheel is decided

(i) to ensure little loss due to fricition

(ii) to ensure full utilization of jet water, i.e. no escape without striking the buckets.

The number of buckets is related to jet ratio

$$z = 0.5m + 15 \text{ for } m \text{ from 6 to 35.}$$

The angular pitch

$$\theta = \frac{2\pi}{z} \text{ (radians)}$$

$$= \frac{360}{z} \text{ (degrees)}$$

3.8 DESIGN EXAMPLE OF A PELTON TURBINE

3.8.1 Design a Pelton Turbine for the Following Site Conditions

1. Gross head, $H_g = 400$ m
2. Power to be developed, $P_t = 30$ MW
3. Penstock length, $L = 2$ km.

Note The power to be developed is decided on the basis of site conditions, load factor, cost of energy produced, revenue expected.

1. Determination of water flow rate

$$P_t = \rho Q g H_g \times \eta_{\text{trans}} \times \eta_{\text{overall}}$$

Assume $\eta_{\text{trans}} = 0.90$

$\eta_{\text{overall}} = 0.91$

$$\therefore \quad Q = \frac{P_t}{\rho g H_g \eta_{\text{tran}} \times \eta_{\text{overall}}}$$

$$= \frac{30 \times 10^6}{1000 \times 9.81 \times 400 \times 0.9 \times 0.91} = 9.976 \text{ m}^3/\text{s}$$

Assume $Q = 10 \text{ m}^3/\text{s}$

2. Number of turbines

The best availability of power supply can be ensured by the use of 3 sets of 50% capacity each so that 2 sets operate at best efficiency and third set to meet peak (surplus) loads. In case of breakdown, two sets meet the total load.

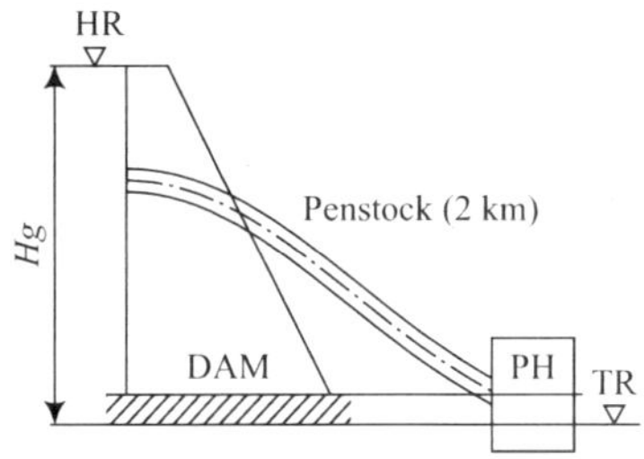

Fig. 3.15 Powerhouse

Therefore, use 3 turbines of 15 MW capacity each.

There will be 3 penstocks handling a discharge rate of 5 m^3/s each.

3. Penstock design

The water flow in a pipe has a velocity range of 1 m/s to 6 m/s. Higher velocity leads to higher pressure drop but smaller diameter of penstock and cost saving.

Assume a suitable water velocity of say 5 m/s. Calculate the pressure drop and operating cost due to power lost in friction. Also calculate the penstock diameter and investment cost on penstock. Plot cost *v/s* pipe diameter and find the pipe diameter for minimum resultant cost. In Fig. 3.15, curve (1) is investment cost *v/s* penstock diameter; curve (2) is operating cost due to power lost in friction *v/s* pipe diameter, curve (3) is the resultant sum of (1) and (2).

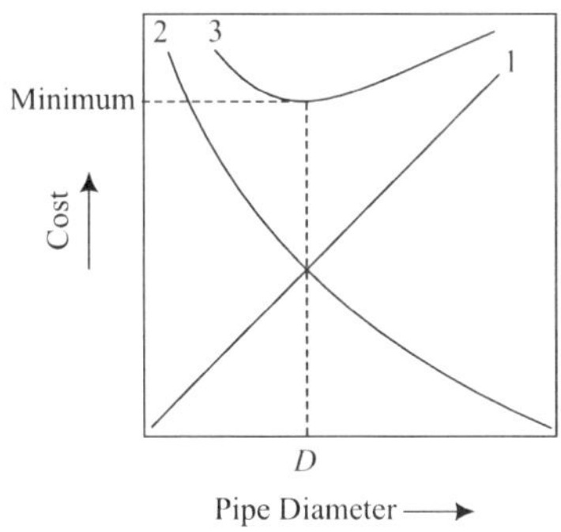

Fig. 3.16 Penstock design

Select diameter D corresponding to minimum total cost.

The cross-sectional area of penstock,

$$A = \frac{\pi}{4}D^2$$

$$A = \frac{Q}{V} = \frac{5}{5} = 1 \text{ m}^2$$

$$\therefore \qquad D = \sqrt{\frac{4A}{\pi}} = \sqrt{\frac{4 \times 1}{\pi}} = 1128 \text{ mm}$$

Refer to Indian standard 1S:6631 and select the nearest size, say 1250 mm.

$$\therefore \qquad V = \frac{Q}{A} = \frac{5}{\frac{\pi}{4}\left(\frac{1250}{1000}\right)^2} = 4 \text{ m/s} \quad \text{OK}$$

As per Darcy's formula, the friction head,

$$\Delta H_f = \frac{4\,fLV^2}{2gD}$$

From curves, $f = 0.007$ (The friction coefficient)

$$\therefore \qquad \Delta H_f = \frac{4 \times 0.007 \times 2000 \times (4)^2}{2 \times 9.81 \times 1.25} = 36.53 \text{ m}$$

$$\eta_{\text{trans}} = \frac{H_g - \Delta H_f}{H_g} = \frac{400 - 36.53}{400} = 0.909 \quad \text{OK}$$

Power available at inlet to turbine

$$p_a = \rho Q g (H_g - \Delta H_f) = 1000 \times 9.81 \times 5(400 - 36.53)$$
$$= 17.85 \text{ MW}$$

4. Selection of turbine speed and type of turbine

It is a high head power plant, Pelton turbine should be most suitable for a head of 400 m.

Selection of turbine speed is based on techno-economical analysis. Higher rotation speed provides:

(i) Compact and economical turbines
(ii) More jet ratio and higher turbine efficiency.
(iii) More generator speed, less number of poles and less cost.
(iv) Higher stresses due to centrifugal force requiring superior materials of construction and more cost of turbine and generator
(v) Higher specific speed and increased turbine efficiency

$$\text{The rotor speed} = \frac{60f}{p}$$

where f = frequency of power generation
= 50 cps for India
p = No. of pair of poles of hydrogenerator

The relation between synchronous speed and the number of pair of poles is shown in Table 3.3.

Table 3.3 Synchronous speed

p	1	2	3	4	5	6	8	10	12
N (rpm)	3000	1500	1000	750	600	500	375	300	250

Assume $N = 500$ rpm

$$\therefore \quad N_S = \frac{N\sqrt{P_t}}{H_a^{5/4}} = \frac{500\sqrt{15{,}000}}{364^{5/4}}$$

$$= 38 \text{ rpm}$$

For Pelton wheel, N_S should be between 10 and 35.

$\therefore$ Use $N = 300$ rpm

$$\therefore \quad N_S = \frac{300\sqrt{15{,}000}}{364^{5/4}} = 22.8 \text{ rpm} \quad \text{OK}$$

$\because$ $N_S < 35$, single jet Pelton turbine can be used.

5. Design of nozzle

The design of nozzle consists of determination of geometry of nozzle and spear and its efficiency.

$$Q = \frac{\pi}{4} d_j^2 V_j \quad [\text{m}^3/\text{s}]$$

Also $V_j = C_v\sqrt{2gHa}$

where coefficient of velocity C_v lies between 0.98 and 0.99.

Assume $C_v = 0.985$

$$\therefore \quad V_j = 0.985\sqrt{2 \times 9.81 \times 364}$$

$$= 83.2 \text{ m/s} \quad \text{OK} \quad [\text{Recommended range 80–85 m/s}]$$

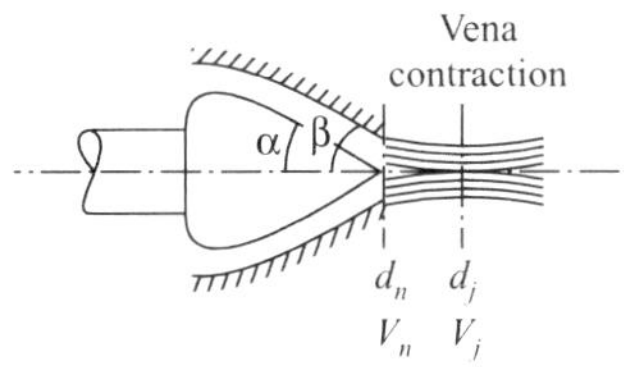

Fig. 3.17 Nozzle design

$$d_j = \sqrt{\frac{4Q}{\pi V_j}} = \sqrt{\frac{4 \times 5}{\pi \times 83.2}} = 0.2766 \text{ m}$$

$$= 276.6 \text{ mm}$$

$$Q = \frac{\pi}{4} d_n^2 V_n$$

$$V_n = C_d \sqrt{2gHa}$$

The coefficient of discharge C_d lies between 0.81 and 0.83.

Take $\quad C_d = 0.82$

$\therefore \quad V_n = 0.82 \sqrt{2 \times 9.81 \times 364}$

$$= 6.93 \text{ m/s}$$

$$d_n = \sqrt{\frac{4Q}{\pi V_n}} = \sqrt{\frac{4 \times 5}{\pi \times 69.3}} = 0.303 \text{ m}$$

$$= 300 \text{ mm}$$

Angle of spear, $\quad 2\alpha = 45° - 50°$

Take $\quad \alpha = 24°$

Angle of nozzle, $\quad 2\beta = 57° - 84°$

Take $\quad \beta = 35°$

Jet power, $\quad P_j = \rho Q g \frac{V_j^2}{2g} = \frac{1000 \times 5}{2} (83.2)^2 \times 10^{-6}$

$$= 17.3 \text{ MW}$$

Power loss in nozzle $= P_a - P_J = 17.85 - 17.3 = 0.55$ MW

Nozzle efficency,

$$\eta_{\text{nozzle}} = \frac{17.3}{17.85} = 0.969$$

$$\approx 97\%$$

6. Runner diameter

The ideal speed ratio, $\frac{V_b}{V_j} = 0.5$

The blade velocity, $V_b = 0.5\ \sqrt{2gHa} = 0.5\ \sqrt{2 \times 9.81 \times 364}$

$= 42$ m/s

$$V_b = \frac{\pi D_m N}{60}$$

The mean diameter of runner,

$$D_n = \frac{60V_b}{\pi N} = \frac{60 \times 42}{\pi \times 300} = 2.6738 \text{ m}$$

$= \mathbf{2675\ mm}$

The jet ratio, $m = \dfrac{D_m}{d_j} = \dfrac{2675}{276.6} = 9.67$

Jet ratio is an important feature of Pelton turbine and influences its characteristics m should be taken from 6.5 to 20.

For maximum efficiency m should be taken from 10 to 14.

Assume $m = 10$

$\therefore$ $D_m = m\,dj = 10 \times 276.6 = \mathbf{2766\ mm}$

7. Number of buckets

The full volume of jet water does not strike the buckets and some water may slip and fall directly into the tail race without doing any useful work. This loss is represented by volumetric efficiency.

$$\eta_{\text{vol}} = \frac{Q - \Delta Q}{Q}$$

The volumetric efficiency ranges between 97 to 99%. It is controlled by number of buckets, shape of bucket and jet ratio (m).

Dr. Taygun's empirical formula for number of buckets

$Z = 0.5$ m $+ 15$

$= 0.5 \times 10 + 15 = 20$

It can be checked by a scale drawing to accommodate the number of buckets calculated. For mechanical stability, number of buckets should be odd.

$\therefore$ Take $Z = \mathbf{21}$

8. Shape of a bucket

The shape of a bucket is related to jet diameter (d_j)

$L = (2.3 \text{ to } 2.8)\ d_j = 636 \text{ to } 774.48$ mm

Take L = **700 mm**

$B = (2.8 \text{ to } 3.2)\ d_j = 774.5 \text{ to } 885 \text{ mm}$

Take B = **830 mm**

$T = (0.6 \text{ to } 0.9)\ d_j = 166 \text{ to } 249 \text{ mm}$

Take T = **200 mm**

Inlet angle, $\beta_1 = 5° \text{ to } 8°$

Take $2\beta_1 = 13°$

Outlet angle, $\beta_2 = 10° \text{ to } 20°$

Take $\beta_2 = 15°$

Notch size = $d_j \approx 300$ mm

Notch depth = $0.2 d_j \approx 50$ mm

Draw the bucket to scale as shown in Fig. 3.18.

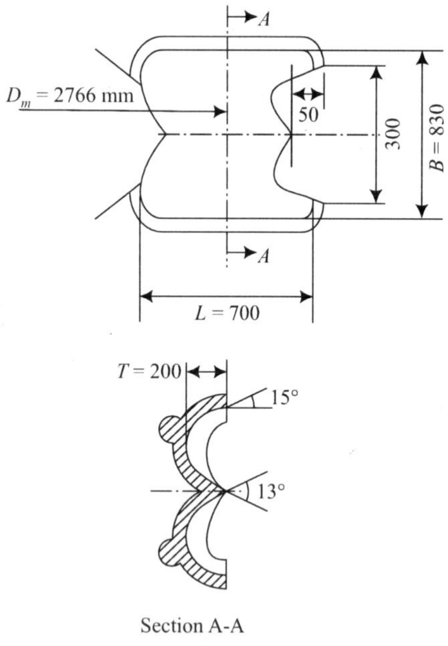

Fig. 3.18 Bucket

3.8.2 Draw Energy Balance of the Pelton Turbine Designed as Per Para 3.7.1

The energy balance sheet is shown in Table 3.4 along with various calculations.

Table 3.4 Energy balance sheet

Basis MW

Credit	*MW*	*Debit*	*MW*
1. Gross power $P_g = \rho Q g H_g \times 10^{-6}$ $= 1000 \times 5 \times 9.81 \times 400 \times 10^{-6}$ 2. Power available $P_a = \rho Q g Ha \times 10^{-6}$	 19.62 17.85	1. Transmission losses $\Delta h_f = P_g - P_a = (19.62 - 17.85)$ 2. Shaft power, P_s 3. Nozzle loses $\Delta h_n = (P_a - P_j) = (17.85 - 17.30)$ 4. Volumetric losses $\Delta h_v = (1 - \eta_v)P_j = (1 - 0.97)17.30$ 5. Hydraulic losses $\Delta h_h = (1 - \eta_h)\eta_v P_j$ $= (1 - 0.92)0.97 \times 17.30$ 6. Mechanical losses $\Delta h_m = (1 - \eta_m)\eta_v \eta_h P_j$	 1.77 15.00 0.55 0.52 1.28 0.55
Total	19.62		19.62

Note: $\eta_{trans} = 0.91$

$\eta_h = 0.92$

$\eta_v = 0.97$

$\eta_m = 0.97$

$$\eta_{overall} = \frac{P_s}{P_a} = \frac{15}{17.85} = \mathbf{0.84 \text{ or } 84\%}$$

QUESTION BANK NO. 3

1. Compare the main hydraulic parameters of Pelton, Francis and Kaplan turbines.
2. What parameters are used for selection of a hydro turbine?
3. Describe the construction and operation of a Pelton turbine. Highlight the role of protection system, speed control system and brakes.
4. Discuss the general arrangement of a Pelton turbine with the help of suitable diagrams.
5. Draw the inlet and outlet velocity triangles for a Pelton turbine.
6. How are work output and hydraulic efficiency calculated from velocity diagrams?
7. What is jet ratio and speed ratio? How are these values calculated for maximum blade efficiency?
8. Draw a Sankey diagram for a Pelton turbine? How are various losses and efficiencies calculated?

9. How is energy balance sheet prepared for a Pelton turbine? Explain the method for calculation of various losses.
10. Explain step-by-step the method for the design of a Pelton turbine.
11. Discuss the role of the following in the design of a Pelton turbine.
    (a) Specific speed
    (b) Jet ratio
12. How are the following dimensions fixed up for a Pelton turbine
    (a) Nozzle diameter
    (b) Runner diameter
    (c) Number of buckets
    (d) Shape of spear and bucket.
13. How are the following features of a Pelton turbine decided?
    (a) Number of jets
    (b) Rotor speed
14. Draw the shape of the bucket of a Pelton turbine. How are various dimensions and angles decided?

TUTORIAL SHEET NO. 3

1. A Pelton wheel develops 3310 kW under a head of 100 m with an overall efficiency of 80%. Find the diameter of the nozzle if the coefficient of velocity for nozzle is 1.

Solution $P_t = 3310$ kW

$$H = 100 \text{ m}$$

$$\eta_{\text{overall}} = 80\%$$

$$P_t = \rho Q g H \, \eta_{\text{overall}}$$

$$\therefore \quad Q = \frac{P_t}{\rho g H \eta_{\text{overall}}} = \frac{3310 \times 10^3}{1000 \times 9.81 \times 100 \times 0.8} = 4.22 \text{ m}^3/\text{s}$$

The velocity of water through nozzle,

$$V_n = C_v \sqrt{2gH} = 1 \times \sqrt{2 \times 9.81 \times 100} = 44.3 \text{ m}^3/\text{s}$$

$$Q = \frac{\pi}{4} d_n^2 \, V_n$$

$$\therefore \quad d_n = \sqrt{\frac{4Q}{\pi V_n}} = \sqrt{\frac{4 \times 4.22}{\pi \times 44.3}} = 0.348 \text{ m}$$

$= \mathbf{348 \text{ mm}}$ Ans.

2. A Pelton wheel is working under a head of 45 m and the rate of flow of water through the jet is 850 litres/sec. The mean bucket speed of the wheel is 15 m/s. Find the efficiency and power produced by the wheel. If the jet is deflected by the bucket through an angle of 165°, draw velocity triangles. Take C_v as 0.98.

Solution $H = 45$ m

$Q = 850$ l/s $= 0.85$ m^3/s

$V_b = 15$ m/s

$V_j = C_v\sqrt{2gH}$

$\therefore \quad = 0.98\sqrt{2\times 9.81\times 45} = 29$ m/s

$$Q = \frac{\pi}{4} d_j^2 V_j$$

$$\therefore \quad d_j = \sqrt{\frac{4Q}{\pi V_j}} = \sqrt{\frac{4\times 0.85}{\pi\times 29}} = 0.193 \text{ m}$$

$= 193$ mm

Speed ratio, $\phi = \dfrac{V_b}{V_j} = \dfrac{15}{29} = 0.517$

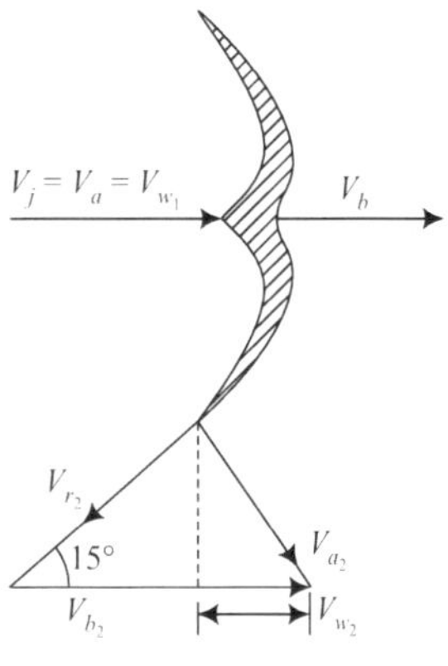

Inlet velocity triangle

$V_j = V_a = V_{w_1} = 29$ m/s

$V_b = 15$ m/s

$V_{r_1} = V_j - V_b = 29 - 15$

$= 14$ m/s

Assume no friction at bucket

$\therefore \quad V_{r_2} = V_{r_1} = 14$ m/s

$V_{w_2} = V_{b_2} - V_{r_2}\cos 15° = 15 - 14\cos 15° = 1.477$ m/s

$P_t = \rho Q(V_{w1} - V_{w2})V_b$

$= 1000\times 0.85(29 - 1.477)\, 15\times 10^{-3}$

$=$ **350.92 kW** Ans

$P_a = \rho QgHa = 1000\times 9.81\times 0.85\times 45\times 10^{-3} = 375.23$ kW

$$\eta_t = \frac{P_t}{P_a} = \frac{350.92}{375.23} = 0.9352$$

$=$ **93.52%** Ans

3. Calculate the number of jets required and their diameter for a Pelton wheel which develops 11,500 kW under a head of 400 m when running at a speed of 450 rpm, assuming that the jet diameter is not to exceed one-tenth of the wheel diameter.

   Assume coefficient of velocity of jet as 0.97 and the ratio of peripheral velocity to the jet velocity as 0.45 and the efficiency of the turbine as 85%. Also find out the number and size of the buckets mounted on the wheel.

   Solution *Data given:*

$$P_t = 11{,}500 \text{ kW}$$

$$H = 400 \text{ m}$$

$$N = 450 \text{ rpm}$$

$$C_v = 0.97$$

$$\frac{V_b}{V_j} = 0.45$$

$$\eta_t = 0.85$$

$$P_t = \rho Q g H \eta_t$$

$$\therefore \quad Q = \frac{P_t}{\rho g H \eta_t} = \frac{11{,}500 \times 10^3}{1000 \times 9.81 \times 400 \times 0.85}$$

$$= 3.44786 \text{ m}^3/\text{s}$$

$$N_S = \frac{N\sqrt{P_t}}{H^{5/4}} = \frac{450\sqrt{11500}}{400^{5/4}} = 26.97 \text{ rpm}$$

$$V_b = 0.45\sqrt{2gH} = 0.45\sqrt{2 \times 9.81 \times 400}$$

$$= 39.865 \text{ m/s}$$

$$V_j = \frac{V_b}{0.45} = 88.59 \text{ m/s}$$

$$V_b = \frac{\pi D_m N}{60}$$

$$\therefore \quad D_m = \frac{60 \times V_b}{\pi N} = \frac{60 \times 39.865}{\pi \times 450} = 1.6919 \text{ m}$$

$$= 1692 \text{ mm}$$

$$m = \frac{D_m}{d_j} = 10$$

$$\therefore \qquad d_j = \frac{D_m}{10} = 169.2 \text{ mm} = 0.169 \text{ m}$$

$$Q_j = \frac{\pi}{4} d_j^2 V_j = \frac{\pi}{4}(0.169)^2 \times 88.59 = 1.9872 \text{ m}^3/\text{s}$$

$$\text{No. of jets} = \frac{Q}{Q_j} = \frac{3.44786}{1.9872} = 1.735 \text{ say } 2$$

$$Q_j = \frac{Q}{2} = \frac{3.44786}{2} = 1.724 \text{ m}^3/\text{s}$$

$$V_j = C_v\sqrt{2gH} = 0.97\sqrt{2 \times 9.81 \times 400} = 85.93 \text{ m/s}$$

$$Q_j = \frac{\pi}{4} d_j^2 V_j$$

$$\therefore \qquad d_j = \sqrt{\frac{4Q_j}{\pi V_j}} = \sqrt{\frac{4 \times 1.724}{\pi \times 85.93}} = \textbf{160 mm}$$

Jet ratio, $$m = \frac{D_m}{d_j} = \frac{1692}{160} = 10.575 \text{ say } 10$$

$$Z = 0.5 \text{ m} + 15$$
$$= 10.575 \times 0.5 + 15 = 20.28 \text{ say } \textbf{21}$$

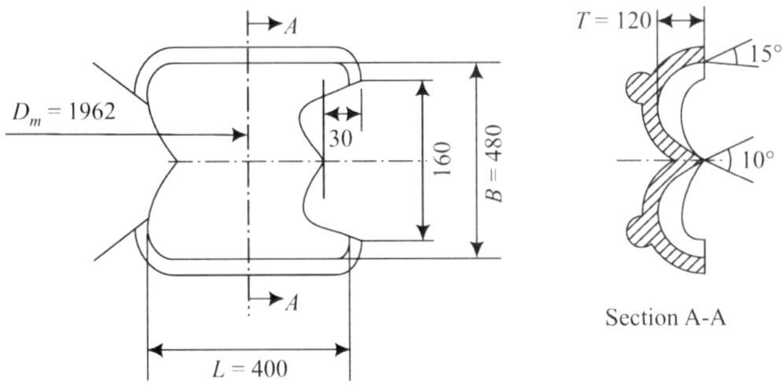

Section A-A

$L = (2.3 - 2.8)d_j = 368$ mm to 448 mm. Take $L = 400$ mm

$B = (2.8 - 3.2)d_j = 448$ mm to 512 mm. Take $B = 480$ mm

$T = (0.6 - 0.9)d_j = 96$ mm to 144 mm. Take $T = 120$ mm

4. Design a Pelton wheel with the following data:

   Head = 200 m

   Power to be developed = 1000 kW

   Speed = 380 rpm

   Assume reasonably all the missing data.

   Solution Follow 3.7.1.

5. (a) How are water turbines classified?

   (b) Why is the jet deflected by the bucket 160° to 165° instead of 180°?

   (c) Sketch the layout of single-jet, double-jet and four-jet Pelton turbines.

6. Show that in a Pelton wheel, where the buckets deflect the water through the angle (180° – α), the hydraulic efficiency of the wheel is given by

$$\eta_h = \frac{2V_b(V_j - V_b)(1 + \cos\alpha)}{V_j^2}$$

where V_j = velocity of the jet

V_b = velocity of the wheel at pitch radius

Solution When a jet of water strikes a series of flat plates mounted on the periphery of a wheel, the efficiency,

$$\eta_h = 2\frac{(V_j - V_b)V_b}{V_j^2}.$$

When a series of curved vanes are mounted on the periphery of a wheel, all the water is utilized since the jet is always intercepted by the vane; the force acting on the vane.

$$F_x = \rho A V_j\,[(V_j - V_b) - \{-(V_j - V_b)\cos\alpha\}]$$

where α = 180° – β

Work done,

$$W = F_x.V_b = \rho A_{Vj}\,(V_j - V_b)(1 + \cos\alpha)$$

The hydraulic efficiency,

$$\eta_h = \frac{W}{\frac{1}{2}(AV_j)V_j^2} = \frac{2(1 + \cos\alpha)(V_j - V_b)V_b}{V_j^2}$$

CHAPTER 4

Francis Turbine

4.1 INTRODUCTION

Francis turbine is a reaction turbine having mixed flow. It has radial entry and exial axit of water. It is a versatile machine and works with medium heads and discharges.

The water from the penstock is admitted to the spiral casing or scroll casing which distributes the water equally around the turbine. The guide vanes (wicket gates) fixed all around the turbine directs the water to the runner blades. A part of pressure energy is converted into kinetic energy in the guide vanes. The running blades utilize the kinetic energy as well as pressure energy of water to produce power. As the runner rotates under the reaction of pressure, it is called reaction turbine.

The whole of pressure energy of water is utilized and water is discharged into a draft tube. The free end of draft tube is submerged deep in the tail water and a closed water passage is obtained. Some kinetic energy of water leaving the turbine is recovered in the draft tube.

Main hydroelectric plants of India use Francis turbine. The examples are Bhakra Dam Project, Hirakund Dam Project, Rihand Dam Project, Cauvery Hydroelectric scheme, Chambal Hydroelectric scheme, etc.

4.2 SPECIFIC SPEED AND TURBINE TYPE

The specific speed of a hydraulic turbine is defined as the speed of a geometrically similar turbine which when operating under a head of 1 m develops 1 kW.

$$N_S = \frac{N\sqrt{P_t}}{H^{5/4}}$$

where P_t = turbine power (kW)

H = water head (m)

N = rotor speed (rpm)

N_S = specific speed (rpm)

The specific speed has been shown to be a useful parameter in classifying hydraulic turbines and in giving information on the general shape of machine.

If performance characteristic, efficiency is plotted against N_S, various types of runners show optimum efficiency in the range of different specific speeds. Hence, different types of runners can be classified according to the range of specific speed. It is, therefore, called "*type characteristic*" or "*character speed*".

Specific speeds of various runners are shown in Fig. 4.1 and Table 4.1

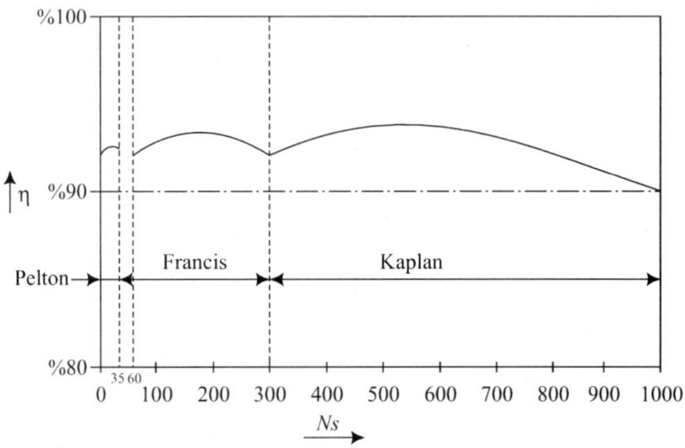

Fig. 4.1 Characteristic speed

Table 4.1 Specific speeds of Various Runners

Turbine	*Type*	N_S
Pelton	Slow	10-20
	Normal	20-28
	Fast	28-35
Francis	Slow	60-120
	Normal	120-180
	Fast	180-300
Kaplan	–	300-1000

While designing a runner for a given output and head, the specific speed can be varied by changing the speed of rotation. It is a common practice to select a high specific speed runner which is always economical because the size of the generator as well as that of power house will become smaller.

High specific speed is essential where water head is low and output is large, because otherwise the speed of rotation will be very low which means cost of generator and power house will be high. On the other hand, there is practically no need of selecting a high value of specific speed for high head installations because even with low specific speed, high rotational speed can be attained with medium capacity plant. High specific speed means greater value of cavitation factor.

As the head for a site is fixed, the specific speed can also be varied by selecting the unit size and number of machines. The specific speed can be changed by changing the runner diameter, width of blades, number of vanes and inlet and outlet angles. For low N_S, large diameter and small tip width are needed and for high N_S, small diameter and large blade width are required.

4.3 MAIN COMPONENTS

The main components of a Francis turbine are:

1. Penstock
2. Spiral casing or scroll casing
3. Guide mechanism
4. Runner and shaft assembly
5. Draft tube

4.3.1 Penstock

It is a water way to connect water reservoir to turbine casing. Water is supplied under pressure, therefore, the penstock is usually fabricated from high quality steel plates. The weld seams are radiographed and subjected to hydraulic test at double the working pressure. If the length of the penstock is short, a separate penstock is run from water reservoir to each turbine. For moderate heads and long distances, a single penstock is used to feed two or more turbines. The penstocks may be buried or embedded and exposed. Expansion joints are used with rigid penstocks having anchorage.

4.3.2 Spiral Casing or Scroll Casing

A spiral casing is provided to receive water from penstock and distribute the same evenly around the turbine to avoid eddies and to ensure equal velocity of water at all sections. It is of varying cross-section and assumes the shape of a spiral. It may be made of concrete for low heads but for high heads, it is fabricated from steel plates or cast steel casings are used.

4.3.3 Guide Mechanism

A series of aerofoil-shaped vanes called the guide vanes or wicket gates, are arranged inside the casing to form a number of flow passages between the casing and runner blades. Aerofoil section ensures minimum frictional and hydrodynamic losses. These vanes can be rotated around their fulcrum and thus the passage are a can be varied. The rotation is received from speed governor which keeps the turbine speed constant with varying loads by adjusting the guide vane water passage and hence the flow. The guide vanes are generally made of cast steel.

4.3.4 Runner and Shaft Assembly

The runner of Francis turbine receives water from guide vanes radially and turns the water through 90° so that water is discharged axially.

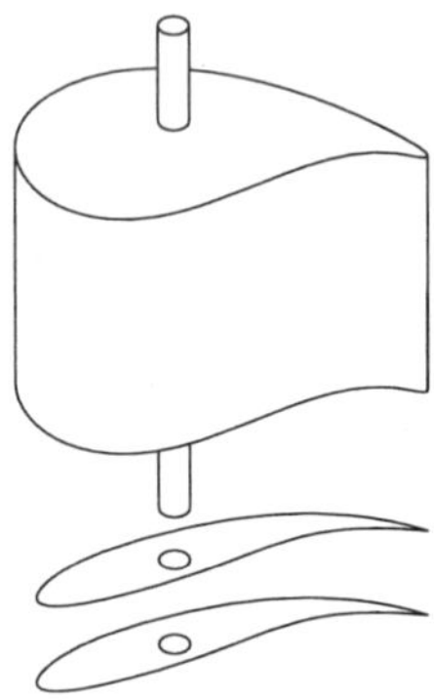

Fig. 4.2 Guide vanes

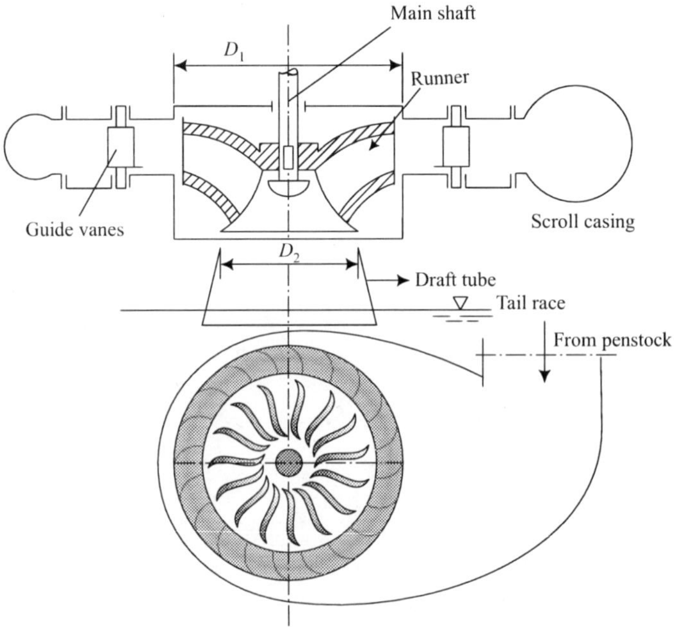

Fig. 4.3 Francis turbine

The shape of the runner depends upon the specific speed. The vanes may be cast integral or fabricated from steel plates. The material of construction is usually cast iron, bronze, cast steel or stainless steel.

The shaft is a forging and is keyed to the runner. The vertical turbines have only one radial-thrust bearing to take the axial load also. The runner works both due to impulsive impact caused by change in direction of flow as well as reactive impact due to change in pressure and velocity. The number of blades may vary between 16 and 24.

4.3.5 Draft Tube

The water from the runner is discharged to the tail race through a gradually expanding pipe called draft tube. The free end is submerged in tail race water. It performs dual function.

1. It converts the kinetic energy of exit water into pressure energy due to increasing cross-sectional area. The exit energy is recovered and turbine efficiency improves.
2. The turbine is installed above the tail race level. There is saving in excavation cost. The maintenance of turbine becomes easier and results in saving in cost.

4.4 VELOCITY TRIANGLES OF FRANCIS TURBINE

In a Francis turbine, the pressure of water is more at the inlet than that at the outlet and water in the turbine flows in a closed conduit. The velocity triangles at inlet and outlet of blades of Francis turbine are shown in Fig. 4.4.

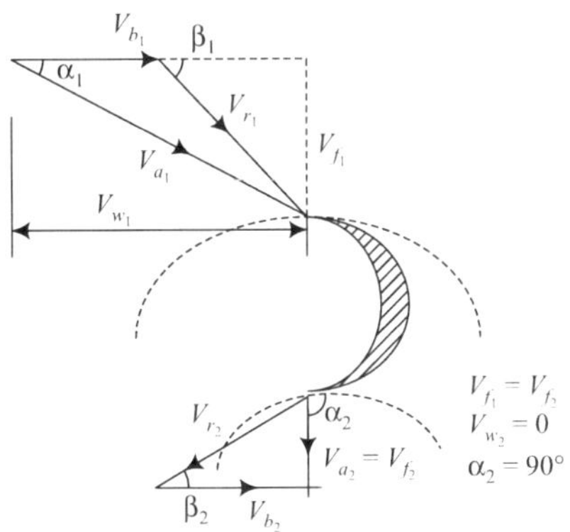

Fig. 4.4 Velocity triangles

The flow velocity remains constant

$$V_{f_1} = V_{f_2}$$

The discharge is axial

$$\therefore \quad V_{w_2} = 0$$

$$\alpha_2 = 0$$

Work done per unit mass flow,

$$W = V_{b_1} V_{w_1}$$

The turbine work,

$$P_t = \rho Q(V_{w_1} V_{b_1} - V_{w_2} V_{b_2})$$

$$\because \quad V_{w_2} = 0$$

$$\therefore \quad P_t = \rho Q V_{w_1} V_{b_1}$$

The inlet velocity triangles depend upon the speed of the runner and are shown in Fig. 4.5.

The outlet velocity triangle is shown in Fig. 4.6.

$$\beta_2 = 90° \pm 30°$$

$$\alpha_2 = 90°$$

$$V_{w_2} = 0$$

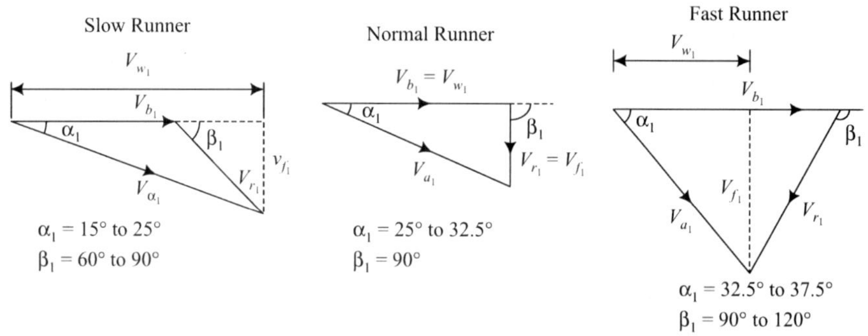

Fig. 4.5 Inlet velocity triangles

There is acceleration of flow in the runner

$$\therefore \quad V_{r_2} > V_{r_1}$$

The hydraulic efficiency,

$$\eta_h = \frac{\rho Q V w_1 V b_1}{\rho Q g H} = \frac{V w_1 V b_1}{g H}$$

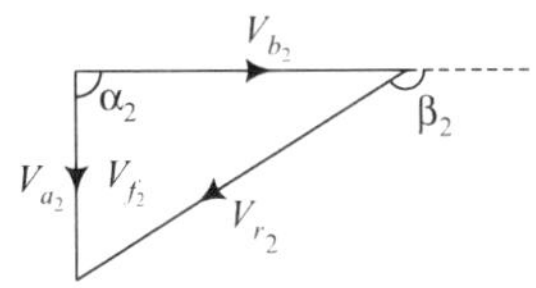

Fig. 4.6 Outlet velocity triangle

$$\tan \alpha_1 = \frac{V_{f_1}}{V_{w_1}}$$

$$\tan \beta_1 = \frac{V_{f_1}}{V_{b_1} - V_{w_1}}$$

$$\tan \beta_2 = \frac{V_{f_2}}{V_{b_2}}$$

$$V_{f_1} = \psi\sqrt{2gH} \quad \text{where} \quad \psi = \frac{V_f}{\sqrt{2gH}} = \text{flow ratio}$$

$$V_{b_1} = \phi\sqrt{2gH} \quad \text{where} \quad \phi = \frac{V_b}{\sqrt{2gH}} = \text{speed ratio}$$

4.5 SANKEY DIAGRAM

The energy flow diagram or Sankey diagram is shown in Fig. 4.7.

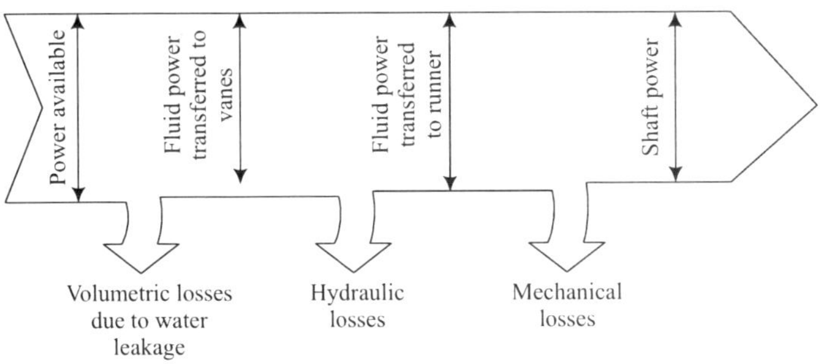

Fig. 4.7 Sankey diagram

Losses

The main losses in a Francis turbine are:

1. Volumetric losses due to leakage of water between guide vanes and runner (ΔQ)
2. Hydraulic losses
   (i) Head loss due to discharge velocity
   (ii) Shock and friction losses over the vanes.
3. Mechanical losses
   (i) Friction losses in the bearings.
   (ii) Windage losses.

Efficiencies

1. Volumetric efficiency,

$$\eta_{vol} = \frac{\text{Fluid Power Transferred to Vanes}}{\text{Power Available at Inlet to Tubine}}$$

$$= \frac{Q - \Delta Q}{Q} \qquad (0.96 - 0.98)$$

2. Hydraulic efficiency

$$\eta_h = \frac{\text{Fluid Power Transferred to Runner}}{\text{Fluid Power Transferred to Vanes}}$$

$$= \frac{V_{w_1} V_{b_1}}{gHa}$$

3. Mechanical efficiency

$$\eta_{\text{mech}} = \frac{\text{Shaft Power}}{\text{Fluid Power Transferred to Runner}} \qquad (0.98 - 0.99)$$

4. Turbine efficiency

   The overall efficiency of turbine

$$\eta_t = \eta_{\text{vol}} \eta_h \eta_{\text{mech}}.$$

4.6 DESIGN OF FRANCIS TURBINE

The site water head and total power to be developed are known, the speed of turbine and turbine power are selected to get specific speed necessary for the type of turbine desired. Practically all dimensions, features and characteristics of the desired turbine can be expressed as a function of the specific speed. Hence, specific speed is the logical key to the design of the turbine.

4.6.1 Speed Ratio and Runner Diameter

The speed ratio,

$$\phi = \frac{V_b}{\sqrt{2gH}}$$

where V_b = Blade peripheral velocity

H = Water head available at turbine inlet

$$\therefore \quad V_b = \phi\sqrt{2gH}$$

The speed ratio which gives maximum turbine efficiency is denoted as ϕ_e and its value ranges between 0.55 and 0.9. This value may be changed in the design by altering certain angles and areas of runner. The recommended values of ϕ_e for different specific speed N_S are given in Fig. 4.8.

$$V_b = \frac{\pi DN}{60}$$

∴ Runner diameter,

$$D = \frac{60\phi_e}{\pi N}\sqrt{2gH}$$

If B is the width or height of the runner, find the ratio B/D from Fig. 4.8.

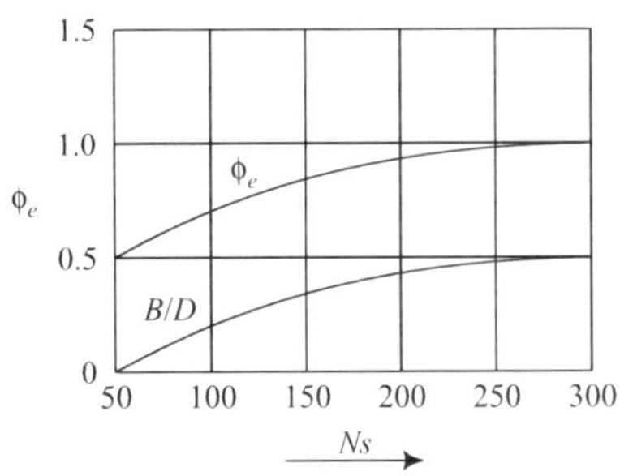

Fig. 4.8 Speed ratio

4.6.2 Flow Ratio

The flow ratio,

$$\psi = \frac{V_f}{\sqrt{2gH}}$$

where V_f = Flow velocity.

The water discharge through turbine,

$$Q = \text{Area of runner} \times \text{Flow velocity}$$

$$= \pi DBV_f$$

Taking into account the effect of blade thickness (about 5%)

$$Q = 0.95\pi DBV_f.$$

$$= 0.95\pi DB\psi\sqrt{2gH}$$

Let $$\frac{B}{D} = m$$

$$\therefore \quad Q = (0.95\pi mD^2)\,\psi\sqrt{2gH} \qquad ...(1)$$

The turbine overall efficiency,

$$\eta_t = \frac{P_t \times 1000}{\rho QgH} \qquad ...(2)$$

where P_t = Turbine power [kW]

The specific speed,

$$N_S = \frac{N\sqrt{P_t}}{H^{5/4}} \qquad ...(3)$$

The flow ratio ψ can be expressed in terms of N_S, ϕ, D and η_t with the help of equations (1), (2), and (3) ψ is useful in fixing inlet and exit angles of blades.

4.6.3 Blade Angles

The whirl ratio,

$$\xi = \frac{V_w}{\sqrt{2gH}} = \frac{\eta_h}{2\phi}$$

where V_w = whirl velocity

$$\tan \alpha_1 = \frac{\psi}{\xi}$$

$$\tan \beta_1 = \frac{\psi}{(\xi - \phi)}$$

$$\alpha_2 = 90°.$$

$$\frac{D_2}{D_1} = 0.5$$

4.6.4 Number of Runner Vanes

In order to avoid any pulsations, the runner vanes are often made an odd number.

Zowski recommendations

Number of runner vanes,

$$n = K\sqrt{D}$$

$K = 3.7$ for slow runner

$K = 3.0$ for normal runner

$K = 2.2$ for fast runner

D = Runner diameter in feet.

4.6.5 Number of Guide Vanes

Zowski recommendations

$$n' = K'\sqrt{D}$$

$K' = 2.5$ for α_1 between 10° and 20°

$K' = 3.0$ for α_1 between 20° and 30°

$K' = 3.5$ for α_1 between 30° and 40°

The values of n and n' should not be same nor of any single multiple to avoid pulsations.

4.7 DESIGN PROBLEM

Design a turbine for the following site conditions.

Head = 120 m

Power = 5 × 120 MW.

No. of penstocks = 5 with head loss due to friction as 3% of gross head.

Length of penstock = 1300 m each.

Solution

1. Determination of water flow rate

Assume overall turbine efficiency, $\eta_t = 0.87$

The friction factor $f = 0.005$

$$\eta_{\text{trans}} = 0.97 \text{ (Given)}$$

Gross head, $H_g = 120$ m (Given)

The turbine power,

$$P_t = \rho Q g H_g \eta_{\text{trans}} \times \eta_t$$

$$\therefore \quad Q = \frac{P_t}{\rho g H g \eta_{trans} \times \eta_t} = \frac{120 \times 10^6}{1000 \times 9.81 \times 120 \times 0.97 \times 0.87}$$

$$= 120 \text{ m}^3/\text{s}.$$

$$\Delta H_f = 0.03 \times 120 = 3.6 \text{ m}$$

$$Q = \frac{\pi}{4} D^2 V$$

$$\therefore \quad V = \frac{4Q}{\pi D^2}$$

$$\Delta H_f = \frac{4 f L V^2}{2gD}$$

$$\therefore \quad 3.6 = \frac{4 \times 0.005 \times 1300}{2 \times 9.81 \times D} = \left(\frac{4 \times 120}{\pi D^2}\right)^2$$

$$D^5 = \frac{4 \times 0.005 \times 1300 \times 120^2 \times 4^2}{3.6 \times \pi^2 \times 2 \times 9.81} = 8593 \text{ m}^5$$

$$D = 6.12 \text{ m say } \mathbf{6 \text{ m}}$$

Water velocity through each penstock,

$$V = \frac{4Q}{\pi D^2} = \frac{4 \times 120}{\pi \times 6^2} = \mathbf{4.25 \text{ m/s}} \quad \text{OK}$$

2. *Type of turbine*

The specific speed,

$$N_S = \frac{N\sqrt{P_t}}{H^{5/4}}$$

From Francis turbine at Bhakra Power Project, $p = 18$ and $N = 166.7$ rpm

$$\therefore \quad N_S = \frac{166.7 \times \sqrt{120 \times 10^3}}{(H_g - \Delta H_f)^{5/4}}$$

$$= \frac{166.7 \times \sqrt{120 \times 10^3}}{116.4^{5/4}}$$

$$= 151 \text{ rpm}.$$

For Francis turbine from Fig. 4.1 and Table 4.1 *normal runner* is suitable for $N_S = 151$ rpm.

3. Spiral casing

The spiral casing diameter d_c can be taken less or equal to penstock diameter. Allowable V_C can be taken as 10 m/s.

$$Q = \frac{\pi}{4} d_c^2 V_C$$

$$\therefore \qquad d_c = \sqrt{\frac{4Q}{\pi V_c}}$$

$$= \sqrt{\frac{4 \times 120}{\pi \times 10}}$$

$$= 3.9 \text{ m}$$

R_{C_1} is fixed to accommodate runner and guide vanes.

$$R_{C_2} = R_{C_1} + \frac{\theta}{2\pi} d_c.$$

The dimensions of spiral casing is fixed by drawing to scale.

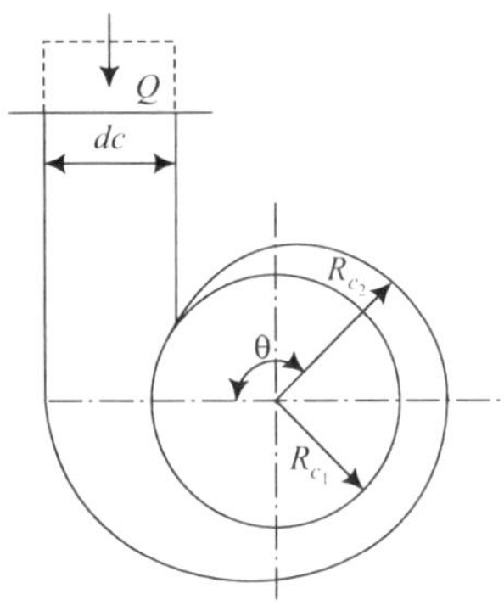

Fig. 4.9 Spiral casing

4. Runner size

The runner diameter,

$$D_1 = \frac{60\phi}{\pi N} \sqrt{2gHa}$$

Read from graph, $\phi = 0.75$

$$\therefore \qquad D_1 = \frac{60 \times 0.75}{\pi \times 166.7} \sqrt{2 \times 9.8 \times 116.4} = 4.1 \text{ m}$$

From graph, $B/D = 0.3$

$\therefore$ $B = 4.1 \times 0.3 = 1.23$ m

5. Velocity triangle

Now, $Q = 0.95 \times \pi B D V_f$

$$V_f = \frac{Q}{0.95 \times \pi \times B \times D} = \frac{120}{0.95 \times \pi \times 1.23 \times 4.1} = 7.97 \text{ m/s.}$$

$$\psi = \frac{V_f}{\sqrt{2gH}} = \frac{7.97}{\sqrt{2 \times 9.81 \times 116.4}} = 0.167$$

Assume $\eta_h = 0.90$

$$\xi = \frac{\eta_h}{2\phi} = \frac{0.9}{2 \times 0.75} = 0.60$$

$$\tan \alpha_1 = \frac{\psi}{\xi} = \frac{0.167}{0.60} = 0.278$$

$\therefore$ $\alpha_1 = 15.5°$

$$\tan \beta_1 = \frac{\psi}{(\xi - \phi)} = \frac{0.167}{(0.60 - 0.75)}$$

$$\beta_1 = -48° = 132°.$$

$$\alpha_2 = 90°$$

6. Number of vanes

(i) Number of runner vanes

$$n = K\sqrt{D}$$

For normal runner, $K = 3.0$

$\therefore$ $n = 3\sqrt{4.1 \times 3.3} = 11$

(ii) Number of guide vanes

$$n' = K'\sqrt{D}$$

$$K' = 2.5 \text{ for } \alpha_1 = 15.5°$$

$\therefore$ $n' = 2.5\sqrt{4.1 \times 3.3} = 9$

QUESTION BANK NO. 4

1. Why specific speed is called "type characteristic"? Classify hydro turbines as per specific speed.
2. What is the role of following components in Francis turbine?
   (a) spiral casing
   (b) draft tube
3. Draw velocity triangles for different Francis runners. How are different angles estimated?
4. List various sources of losses in a Francis turbine. Draw a Sankey diagram and define different efficiencies.
5. Explain step-by-step the design of a Francis turbine.
6. Define the following. Explain their role in the design of a Francis turbine.
   (a) speed ratio
   (b) flow ratio

TUTORIAL SHEET NO. 4

1. An inward flow reaction turbine has outer and inner diameters of the wheel as 1 m and 0.5 m respectively. The vanes are radial at inlet and axial at outlet and water enters the vanes at an angle of 10°. Assuming the velocity of flow as constant and equal to 5 m/s, calculate:
   (a) The speed of wheel, and
   (b) The vane angle at outlet.

Solution The data given:

$$D_1 = 1 \text{ m}$$
$$D_2 = 0.5 \text{ m}$$
$$V_{f_1} = V_{f_2} = 5 \text{ m/s}$$
$$\alpha_1 = 10°$$
$$\beta_1 = 90°$$
$$\alpha_2 = 90°$$

The velocity triangles are drawn below.

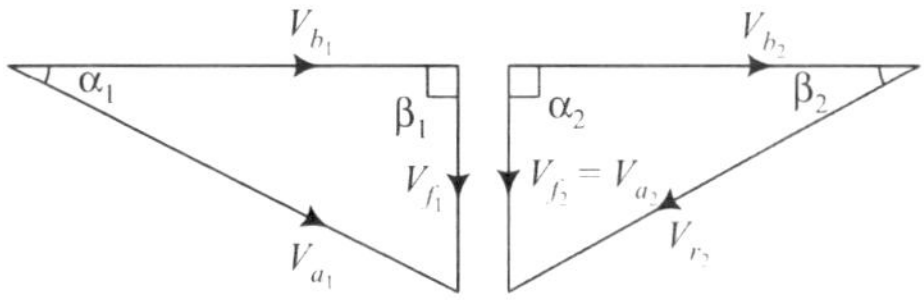

(a) $$\tan \alpha_1 = \frac{V_{f_1}}{V_{b_1}}$$

$\therefore$ $$V_{b_1} = \frac{5}{\tan 10°} = 28.356 \text{ m/s}$$

Now $$V_{b_1} = \frac{\pi D_1 N}{60}$$

$\therefore$ $$N = \frac{60 \times V_{b1}}{\pi D_1} = \frac{60 \times 28.356}{\pi \times 1} = \mathbf{541.56\ m/s} \quad \text{Ans.}$$

(b) $$V_{b_2} = \frac{\pi D_2 N}{60} = \frac{\pi \times 0.5 \times 541.56}{60} = 14.176 \text{ m/s}$$

$$\tan \beta_2 = \frac{V_{f_2}}{V_{b_2}} = \frac{5}{14.176}$$

$\therefore$ $$\beta_2 = \mathbf{19°25'} \quad \text{Ans.}$$

2. An inward flow reaction turbine, having external diameter of 1 m is running at a speed of 180 rpm. The guide blade angle is 15°. If the velocity of flow at inlet is 5 m/s, find:
   (a) peripheral velocity at inlet
   (b) velocity of whirl at inlet
   (c) absolute velocity of water at inlet
   (d) vane angle at inlet

Solution The data given

$$D_1 = 1 \text{ m}$$
$$N = 180 \text{ rpm.}$$
$$\alpha_1 = 15°$$
$$V_{f_1} = 5 \text{ m/s}$$

Draw inlet velocity triangle.

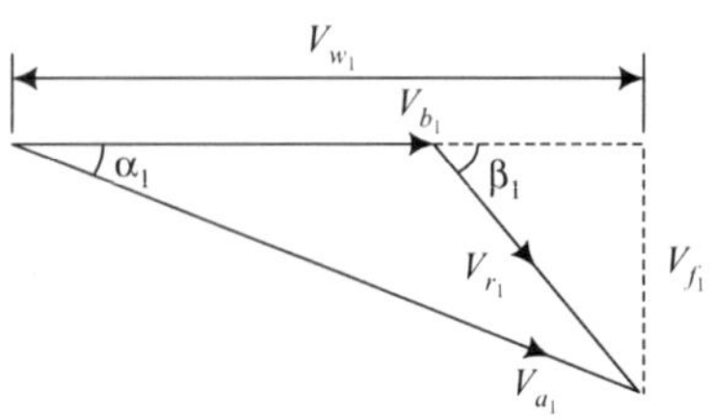

(a) $$V_{b_1} = \frac{\pi D_1 N}{60} = \frac{\pi \times 1 \times 180}{60}$$

$$= \mathbf{9.42\ m/s} \quad \text{Ans}$$

(b) $$\tan \alpha_1 = \frac{V_{f_1}}{V_{w_1}}$$

$$\therefore \quad V_{w_1} = \frac{V_{f_1}}{\tan \alpha_1} = \frac{5}{\tan 15^\circ} = \mathbf{18.66\ m/s} \quad \text{Ans}$$

(c) $$\sin \alpha_1 = \frac{V_{f_1}}{V_{a_1}}$$

$$\therefore \quad V_{a_1} = \frac{V_{f_1}}{\sin \alpha_1} = \frac{5}{\sin 15^\circ} = \mathbf{19.3\ m/s} \quad \text{Ans}$$

(d) $$\tan \beta_1 = \frac{V_{f_1}}{V_{b_1} - V_{w_1}} = \frac{5}{V_{w_1} - V_{b_1}} = \frac{5}{18.66 - 9.42} = 0.541$$

$$\therefore \quad \beta_1 = \mathbf{28.42^\circ} \quad \text{Ans}$$

3. Estimate the main dimensions for an inward flow turbine to suit the following conditions.

Head = 58 m	Hydraulic efficiency = 90%
Power developed = 400 kW	Overall efficiency = 85%
Speed = 1000 rpm	Flow ratio = 0.16

The ratio of wheel width to wheel diameter at inlet = 0.1.

The ratio of inner diameter and outer diameter = 0.5.

Consider constant velocity of flow and axial of discharge.

Solution The data given:

$$H = 58 \text{ m}$$

$$P_t = 400 \text{ kW}$$

$$N = 1000 \text{ rpm}$$

$$\eta_h = 0.90$$

$$\eta_h = 0.85$$

$$\psi = 0.16$$

$$B/D = 0.1$$

$$\frac{D_2}{D_1} = 0.5$$

$$V_{f_1} = V_{f2}$$
$$\alpha_2 = 90°$$

(a) The specific speed,

$$N_S = \frac{N\sqrt{P_t}}{H^{5/4}} = \frac{1000\sqrt{400}}{58^{5/4}} = \frac{1000 \times 20}{160}$$

$$= 125 \text{ rpm} \quad \textit{Normal Francis Runner}$$

$$P_t = \rho Q g H \eta_t$$

$$\therefore \quad Q = \frac{P_t}{\rho g H \eta_t} = \frac{400 \times 10^3}{1000 \times 9.81 \times 58 \times 0.85}$$

$$= 0.827 \text{ m}^3/\text{s}$$

$$V_f = \psi\sqrt{2gH} = 0.16\sqrt{2 \times 9.81 \times 58}$$

$$= 5.4 \text{ m/s}$$

$$Q = \pi D_1 B V_f$$
$$= \pi D_1 (0.1 D_1) V_f$$

$$\therefore \quad D_1 = \sqrt{\frac{Q}{\pi \times 0.1 \times V_f}} = \sqrt{\frac{0.827}{\pi \times 0.1 \times 5.4}}$$

$$\therefore \quad \left.\begin{aligned} D_1 &= 0.6982 \text{ m} \approx \mathbf{0.7\ m} \\ D_2 &= \frac{D_1}{2} = \mathbf{0.35\ m} \\ B &= 0.1 D_1 = \mathbf{0.07\ m} \end{aligned}\right\} \text{Ans}$$

(b)
$$V_{b1} = \frac{\pi D_1 N}{60} = \frac{\pi \times 0.7 \times 1000}{60} = 36.65 \text{ m/s}$$

$$\phi = \frac{V_{b_1}}{\sqrt{2gH}} = \frac{36.65}{\sqrt{2 \times 9.81 \times 58}} = 1.08645$$

$$\xi = \frac{\eta_h}{2\phi} = \frac{0.90}{2 \times 1.08645} = 0.4142$$

$$\tan \alpha_1 = \frac{\psi}{\xi} = \frac{0.16}{0.4142}$$

$$\therefore \quad \alpha_1 = 21°$$

$$\tan \beta_1 = \frac{\psi}{(\xi - \phi)} = \frac{0.16}{(0.4142 - 1.08645)} = -0.238$$

$$\beta_1 = 13.4° \text{ or } \mathbf{166.6°}$$

(c) Number of runner vanes,

$$n = 3\sqrt{D \times 3.3}$$

$$= 3\sqrt{0.7 \times 3.3} = 4.55 \text{ or } 5$$

Number of guide vanes,

$$n' = 2.5\sqrt{D \times 3.3} = 3.8 \text{ or } 4$$

CHAPTER 5

Propeller and Kaplan Turbines

5.1 INTRODUCTION

The Francis turbines are medium head turbines and hydraulic efficiency is affected due to following reasons:

1. The number of blades are more from 16 to 24.
2. The blades receive water in radial direction and water is discharged in axial direction. There is 90° bend to water.

Those two sources of hydraulic losses are overcome in Propeller turbine which have axial entry and axial discharge. The number of blades are only 3 to 6. These are also reaction turbines and work with large flow and low heads. The specific speed is high between 300 and 1000 rpm.

The components of propeller turbine such as scroll casing, guide mechanism and draft tube are similar to Francis turbine except the runner.

The schematic diagram of a propeller turbine is shown in Fig. 5.1. The runner blades are fixed in position as in Francis turbine. The efficiency depends upon inlet blade angle. In fixed blade runners, it is not possible to vary the inlet blade angle for varying loads. The propeller turbine is designed for maximum efficiency only for a particular load. At all other loads, efficiency falls. These turbines are called peak load turbines.

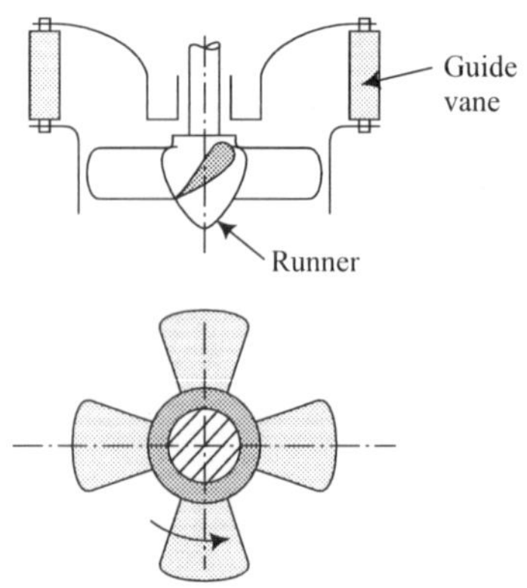

Fig. 5.1 Propeller turbine

5.2 KAPLAN TURBINE

Kaplan turbine is a special propeller turbine in which the individual runner blades are pivoted to the hub, so that their inclination may be adjusted during operation responding to changes in load. The blades are adjusted automatically rotating about pivots with the help of a governor servo-mechanism. The blades are attached to a hollow boss and hollow shaft. The servo-mechanism to rotate the blades are housed inside the hollow boss and shaft. The turbine can be run at maximum efficiency at all loads because the inlet angle is automatically adjusted with variation of load and a flat curve is obtained.

The guide vanes are rotated by speed governor during load fluctuation to ensure constant speed. Therefore, in Kaplan turbine both guide vanes as well as runner blades are rotated by speed governor during load fluctuation.

The arrangement drawing of a Kaplan turbine is shown in Fig. 5.3.

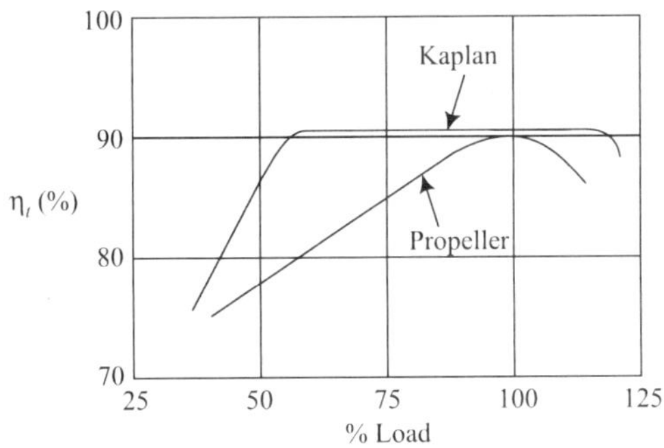

Fig. 5.2 Efficiency curves

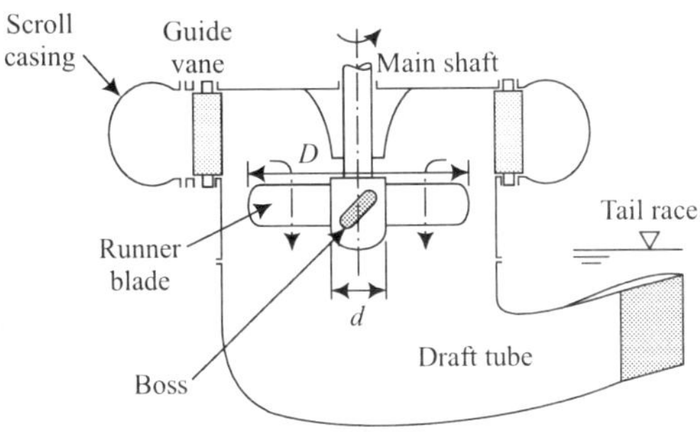

Fig. 5.3 Kaplan turbine

5.3 TURBINE CONSTANTS

5.3.1 Velocity Triangles

$$D = \text{Diameter of runner}$$

$$d = \text{Diameter of boss.}$$

$$\frac{d}{D} = 0.38 + \frac{H}{220}$$

At edge of diameter D,

$$V_{b_1} = V_{b_2}$$

At the hub diameter

$$V'_{b_1} = V'_{b2}$$

$$V'_{f_1} = V'_{f_2} = V_{f_1} = V_{f_2}.$$

$$\alpha_1 = 45° \text{ to } 70°$$

$$\beta_1 = 20° \text{ to } 68°$$

The degree of reaction is less than 50% and increases gradually from the hub to the tip.

1. water inlet
2. water outlet

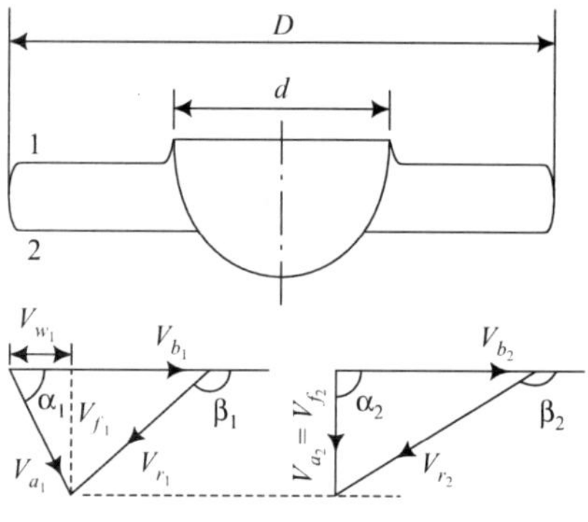

Fig. 5.4 (a) Velocity triangles at edge (D)

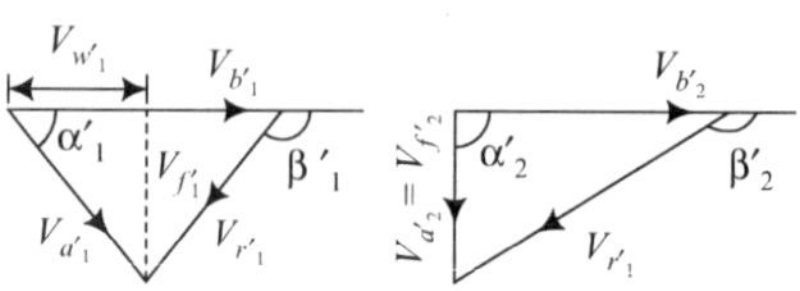

Fig. 5.4 (b) Velocity triangles at hub (d)

5.3.2 Ratios

1. Speed ratio

   The blade speed,

$$V_b = \phi\sqrt{2gH}$$

where ϕ = speed ratio

$$= \frac{V_b}{\sqrt{2gH}}$$

$$V_b = \frac{\pi DN}{60}$$

ϕ is a function of N_S.

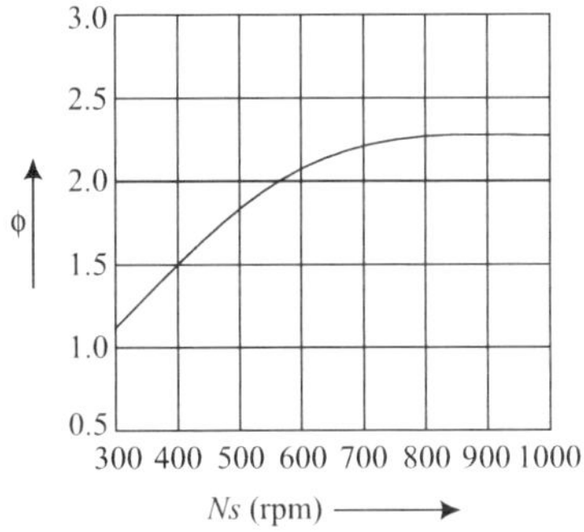

Fig. 5.5 Speed ratio

2. Flow ratio

   The flow velocity

$$V_f = \psi\sqrt{2gH}$$

where ψ = Flow ratio.

$$Q = K\frac{\pi}{4}(D^2 - d^2)\,V_f$$

where K = Coefficient of flow area after deduction of area occupied by the blades.

3. Whirl ratio

   The whirl velocity,

$$V_w = \xi\sqrt{2gH}$$

$\therefore$ Whirl ratio $\quad \xi = \dfrac{V_w}{\sqrt{2gH}} = \dfrac{\eta_h}{2\phi}$

where hydraulic efficiency, $\eta_h = 0.95$.

5.3.3 Blade Angles

$$\tan \alpha_1 = \frac{V_{f_1}}{V_{w_1}} = \frac{\psi\sqrt{2gH}}{\xi\sqrt{2gH}} = \frac{\psi}{\xi}$$

$$\tan \beta_1 = \frac{V_{f_1}}{V_{b_1} - V_{w_1}} = \frac{\psi}{(\phi - \xi)}$$

$$\tan \beta_2 = \frac{V_{f_2}}{V_{b_2}} = \frac{\psi}{\phi}$$

$$V'_{w_1} = V_{w_1} \times \frac{D}{d}$$

$$\therefore \qquad \xi' = \xi\,\frac{D}{d}$$

$$V'_{b_1} = \frac{\pi d N}{60}$$

$$\therefore \qquad \phi' = \phi\frac{D}{d}$$

5.3.4 Losses and Efficiencies

These can be estimated in the same way as for Francis turbine. See Art. 4.5.

5.4 DESIGN PROBLEM

Design a propeller turbine of 8 MW with water head of 5 m.

5.4.1 Specific Speed

$$N_S = \frac{N\sqrt{P_t}}{H^{5/4}}$$

N_S varies from 300 rpm to 1000 rpm.

Take $N_S = 300$ rpm.

$$\therefore \qquad N = \frac{N_s H^{5/4}}{\sqrt{P_t}} = \frac{300 \times 5^{5/4}}{\sqrt{8000}} = 25 \text{ rpm.}$$

Take $N_S = 1000$ rpm.

$$\therefore \qquad N = 25 \times \frac{1000}{300} = 83 \text{ rpm}$$

Select a synchronous speed of 75 rpm which can be coupled to a generator having 40 pairs of poles for a generation frequency of 50 cps.

$$\therefore \qquad N_S = \frac{N\sqrt{P_t}}{H^{5/4}} = \frac{75\sqrt{8000}}{5^{5/4}} = 897 \text{ rpm.}$$

From graph for $N_S = 897$ rpm, speed ratio $\phi = 2.2$.

5.4.2 Water Flow

$$P_t = \rho Q g H \eta_t$$

Assume $\eta_t = 0.90$

$$\therefore \qquad Q = \frac{P_t}{\rho g H \eta_t} = \frac{8 \times 10^6}{1000 \times 9.81 \times 5 \times 0.90} = 181.22 \text{ m}^3\text{/s.}$$

5.4.3 Runner Dimensions

Runner diameter,

$$D = \frac{60\phi}{\pi N}\sqrt{2gH} = \frac{60 \times 2.2}{\pi \times 75}\sqrt{2 \times 9.81 \times 5}$$

$$= 5.55 \text{ m}$$

Hub diameter,

$$d = \left(0.38 + \frac{H}{220}\right) 5.55 = 2.235 \text{ m}$$

5.4.4 Velocity Diagrams

Refer Fig. 5.4.

$$\phi = 2.2,\ \psi = 0.90;\ \xi = \frac{\eta_h}{2\phi} = \frac{0.95}{2 \times 2.2} = 0.216$$

$$\tan \alpha_1 = \frac{\psi}{\xi} = \frac{0.90}{0.216} \qquad \therefore \alpha_1 = \mathbf{76.5^\circ}$$

$$\tan \beta_1 = \frac{\psi}{(\phi - \xi)} = \frac{0.90}{(2.2 - 0.216)} \qquad \therefore \beta_1 = \mathbf{24.4°}$$

$$\tan \beta_2 = \frac{\psi}{\phi} = \frac{0.90}{2.2} \quad \beta_2 = 22.2°$$

5.5 OTHER TURBINES

5.5.1 Deriaz Turbine

Deriaz turbine is a reversible flow turbine and is a cross between Kaplan and Francis turbines. It has 10 to 12 inclined blades which are adjustable like Kaplan turbine. For pumped-storage power plants it works both as turbine as well as pump. For tidal pover plants it can work in either direction of flow during high tide where water is being stored in the reservoir and low tide when water flows from reservoir to the ocean. It is also called diagonal turbine as flow over runner is at angle of 45° to the axis. The general arrangement is shown in Fig. 5.6. These turbines can work even upto head of 200 m.

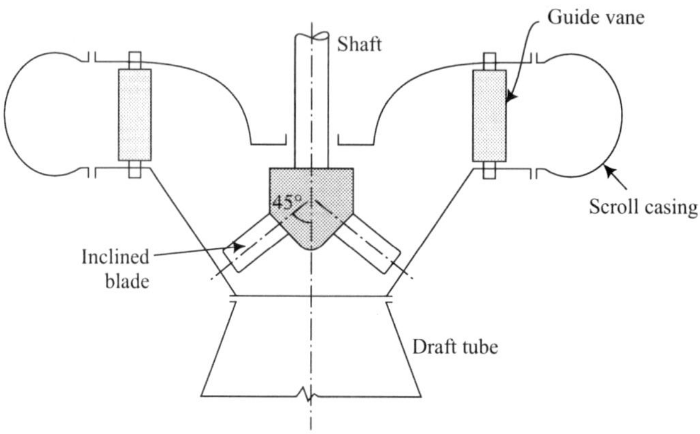

Fig. 5.6 Deriaz turbine

5.5.2 Bulb Turbine

Bulb or tubular turbines are small fixed axial flow propeller turbines operating under low heads. The hydro-generator is enclosed in a bulb-shaped casing and installed in the middle of flow passage. The bulb and propeller form an integral unit. There is a straight conical draft tube.

These are suitable for tidal power plants, run-off the river plants.

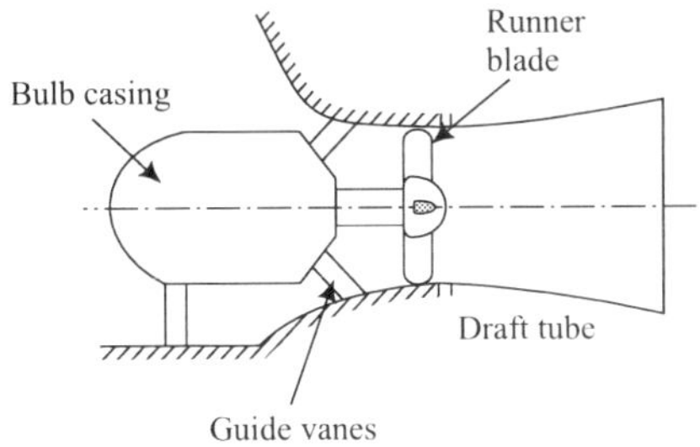

Fig. 5.7 Bulb turbine

QUESTION BANK NO. 5

1. What is the main difference between a Francis and propeller turbine? How are hydraulic losses minimized in propeller turbine?
2. What is a peak load turbine and flat curve turbine? How is a propeller turbine converted into Kaplan turbine?
3. How are blade angle fixed in Kaplan turbine in terms of turbine constants?
4. Write brief notes on the following:
   (a) Deriaz turbine
   (b) Bulb turbine

TUTORIAL SHEET NO. 5

1. A Kaplan turbine develops 10,000 kW under a head of 4.3 m. Find the diameter and speed of the turbine if the speed ratio is 1.8, flow ratio is 0.5, boss diameter as 0.35 times the outer diameter and overall turbine efficiency is 90%.

Solution The given data:

$$P_t = 10{,}000 \text{ kW}$$

$$H = 4.3 \text{ m}$$

$$\phi = 1.8$$

$$\psi = 0.50$$

$$\frac{d}{D} = 0.35$$

$$\eta_t = 90\%$$

$$V_f = \psi\sqrt{2gH}$$

$$= 0.5\sqrt{2 \times 9.81 \times 4.3} = 4.6 \text{ m/s.}$$

$$P_t = \rho Q g H \eta_t$$

$$\therefore \quad Q = \frac{P_t}{\rho g H \eta_t} = \frac{10{,}000 \times 10^3}{1000 \times 9.81 \times 4.3 \times 0.9}$$

$$= 263.4 \text{ m}^3/\text{s}$$

$$Q = \frac{\pi}{4}(D^2 - d^2) V_f$$

$$= \frac{\pi}{4}[D^2 - (0.35D)^2] V_f$$

$$\therefore \quad D = \sqrt{\frac{4Q}{\pi(1 - 0.35^2) V_f}} = \sqrt{\frac{4 \times 263.4}{\pi \times 4.6 \times 0.8775}}$$

$$= 9.1 \text{ m say } \mathbf{9\ m}$$

$$d = 0.35\, D = \mathbf{3.15\ m}$$

$$V_b = \phi\, \sqrt{2gH} = 1.8\, \sqrt{2 \times 9.81 \times 4.3}$$

$$= 16.5 \text{ m/s}$$

$$V_b = \frac{\pi D N}{60}$$

$$\therefore \quad N = \frac{60 \times V_b}{\pi D} = \frac{60 \times 16.5}{\pi \times 9} = \mathbf{35\ rpm}$$

2. Find the specific speed and the type of water turbine developing shaft power of 5145 kW under a head of 20 m when running at 100 rpm. Calculate its normal speed and output under a head of 2.5 m.

   Solution (a) Data given:

$$P_t = 5145 \text{ kW}$$

$$H = 20 \text{ m}$$

$$N = 100 \text{ rpm.}$$

The specific speed,

$$N_S = \frac{N\sqrt{P_t}}{H^{5/4}} = \frac{100\sqrt{5145}}{20^{5/4}}$$

$$= 169.6 \text{ rpm.}$$

It is a Francis turbine.

(b) $$P_t = \rho Q g H \eta_t$$

$$Q = \psi\sqrt{2gH}$$

$$P_t \propto QH$$

$$\propto H^{\frac{3}{2}}$$

$$\frac{P'_t}{P_t} = \left(\frac{H'}{H}\right)^{\frac{3}{2}}$$

$$P'_t = P_t\left(\frac{H'}{H}\right)^{\frac{3}{2}} = 5145\left(\frac{2.5}{20}\right)^{\frac{3}{2}} = 227 \text{ kW.}$$

$$V_b = \phi\sqrt{2gH}$$

$$\frac{2\pi ND}{60} = V_b = \phi\sqrt{2gH}$$

$\therefore$ $$N \propto \sqrt{H}$$

$$\frac{N'}{N} = \left(\frac{H'}{H}\right)^{\frac{1}{2}}$$

$\therefore$ $$N' = N\left(\frac{H'}{H}\right)^{\frac{1}{2}} = 100\left(\frac{2.5}{20}\right)^{\frac{1}{2}} = 35 \text{ rpm.}$$

Also $$N' = \frac{N'_s (H')^{\frac{5}{4}}}{(P'_t)^{\frac{1}{2}}} = \frac{170(2.5)^{\frac{5}{4}}}{\sqrt{227}}$$

$$= 35 \text{ rpm.}$$

3. Calculate the diameter and speed of the runner of a Kaplan turbine developing 6000 kW under an effective head of 5 m. Overall efficiency of the turbine is 90%. The diameter of the boss is 0.4 times the external diameter of runner. The turbine speed ratio is 2.0 and flow ratio is 0.6. What is the specific speed of the turbine?

[D = 5.66 m; N = 66.8 rpm, N_S = 692 rpm].

CHAPTER 6

Draft Tube and Cavitation

6.1 INTRODUCTION

Draft tube is an airtight diverging conduit for carrying water from turbine runner exit to the tail race. The gradually increasing cross-section of draft tube helps to convert the kinetic energy of exit water into pressure energy. It is necessary and an integral part of all reaction turbines (Francis, Propeller, Kaplan, Bulb). The draft tube may be a concrete tunnel or fabricated from steel. The outlet end of draft tube is submerged in the water approximately 1 m below the lowest tail race level.

A draft tube is shown in Fig. 6.1 where section 1-1 is the turbine exit or draft tube inlet and section 2 – 2 is the draft tube outlet located at a depth of Z_2 below tail race level. Apply Bernoulli's equation between section 1 – 1 and 2 – 2,

$$\frac{p_1}{w} + \frac{V_1^2}{2g} + Z_1 = \frac{p_2}{w} + \frac{V_2^2}{2g} + 0 + h_f$$

where h_f = hydraulic loss in draft tube.

$$\therefore \qquad \frac{p_1}{w} = \frac{p_2}{w} - Z_1 - \left(\frac{V_1^2 - V_2^2}{2g} - h_f\right)$$

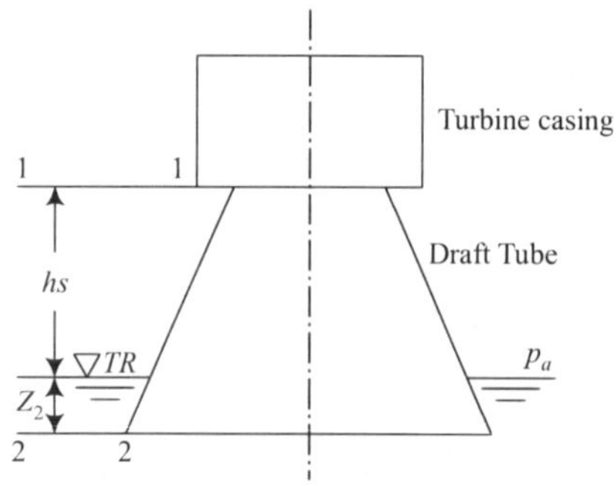

Fig. 6.1 Draft tube

But $$\frac{p_2}{w} = \frac{p_a}{w} + Z_2$$

$$\frac{p_1}{w} = \frac{p_a}{w} + (Z_2 - Z_1) - \left(\frac{V_1^2 - V_2^2}{2g} - h_f\right)$$

$$w = \rho_g$$

$$Z_2 - Z_1 = h_s = \text{suction head of draft tube}$$

$$\frac{V_1^2 - V_2^2}{2g} = \text{dynamic head.}$$

∴ $$\frac{p_1}{w} = \frac{p_a}{w} - h_s - \left(\frac{V_1^2 - V_2^2}{2g} - h_f\right)$$

∴ $$p_1 < p_a.$$

The cross-sectional area of draft tube is increasing.

$$A_2 > A_1$$

∴ $$V_2 < V_1$$

The suction head, h_s is positive, the available head of turbine increases leading to higher turbine output and efficiency. The actual gain of pressure head is $\left(\frac{V_1^2 - V_2^2}{2g} - h_f\right)$.

The efficiency of draft tube

$$\eta_d = \frac{\frac{V_1^2 - V_2^2}{2g} - h_f}{\frac{V_1^2}{2g}}$$

Let $$h_f = k\frac{(V_1^2 - V_2^2)}{2g}$$

∴ $$\eta_d = \frac{(1-k)\left(\frac{V_1^2 - V_2^2}{2g}\right)}{V_1^2/2g}$$

6.2 FUNCTIONS OF DRAFT TUBE

A draft tube performs the following functions.

1. Recovery of exit loss

The water leaves the turbine runner at a high velocity (V_1) and carries away with it a large kinetic energy $\left(\frac{V_1^2}{2g}\right)$ which forms a big loss mechanism. The diffuser action of draft tube recovers this energy by converting kinetic energy $\left(\frac{V_1^2 - V_2^2}{2g}\right)$ into pressure energy (h_s). The water is discharged into tail race at very low velocity (V_2). The pressure in the draft tube is above the vapour pressure corresponding to the water temperature and cavitation is avoided. The effective head of the turbine is increased by ($h_s - h_f$), raising the output and efficiency of turbine.

2. Turbine installation above water level

Draft tube ensures an airtight closed circuit at turbine outlet thereby making possible to decrease the pressure at turbine outlet below atmospheric ($p_1 < p_a$). For same working head, now it is possible to install the turbine above tail race water level. There is saving in excavation cost and maintenance cost. If there was no draft tube, the turbine would be installed below the tail race level and water had to be pumped out for maintenance of the runner. Therefore, draft tube helps to avoid the need of pumping out of water.

6.3 TYPES OF DRAFT TUBES

The draft tubes are designed to achieve the following:

1. To ensure the maximum possible recovery of kinetic energy which is expressed as draft tube efficiency.

$$\eta_d = \frac{\text{Actual head recovered}}{\text{Theoretical head}} = \frac{\text{Theoretical head-losses}}{\text{Theoretical head}}$$

   The losses in the draft tube are:

   (i) Due to separation of flow from walls of the tube, the taper angle is restricted to 7° to 8°, see Fig. 6.3.

   (ii) Due to formation of vortex at the tube exit. This is controlled by provision of vortex breaker. See Fig. 6.4 where a cone is provided as vortex breaker.

   (iii) Friction losses in the tube.

   (iv) Head loss due to tube bends.

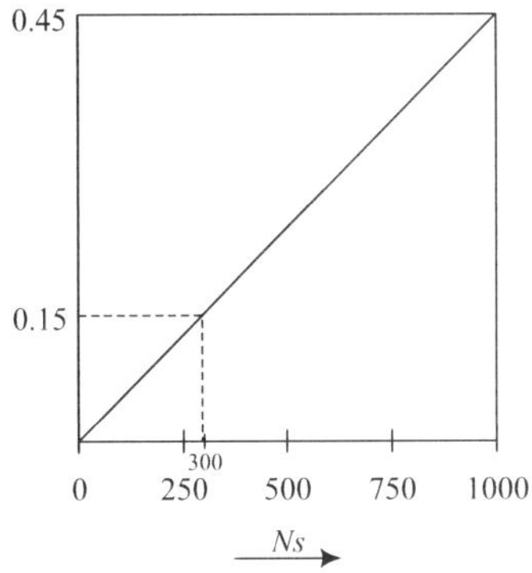

Fig. 6.2 Draft tube efficiency

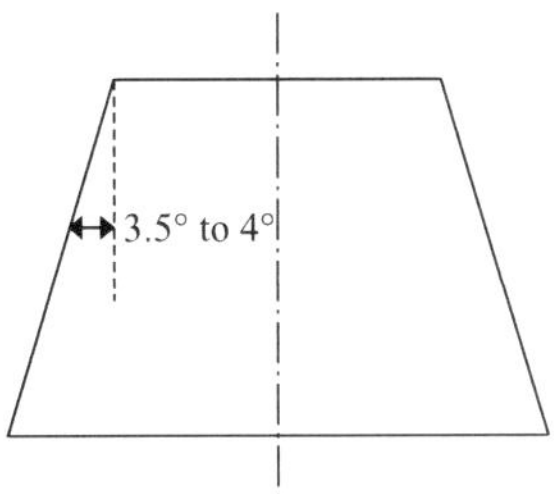

Fig. 6.3 Taper angle

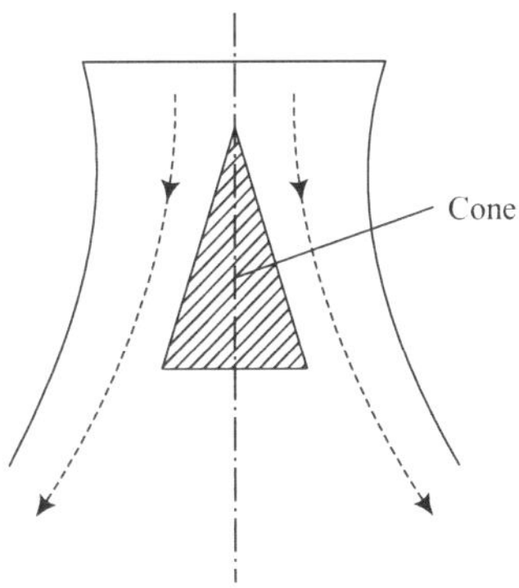

Fig. 6.4 Vortex breaker

2. To ensure minimum cost of excavation to lay down the tube. This is achieved by the use of bent tube.

   The following types of draft tubes are used.

   (i) Straight divergent tube. This gives $\eta_d = 85\%$

   (ii) Bell mouth tube. This gives $\eta_d = 85\%$

   (iii) Hydro cone or Moody's spreading tube, $\eta_d = 85\%$

   (iv) Simple elbow tube. Low excavation but $\eta_d = 66\%$

   (v) Elbow tube with circular inlet and rectangular outlet. This requires minimum depth of excavation and η_d is more than simple elbow.

6.4 CAVITATION

Vapour pressure of water is a function of its temperature. During flow of water in a turbine, whenever, water pressure falls below the vapour pressure corresponding

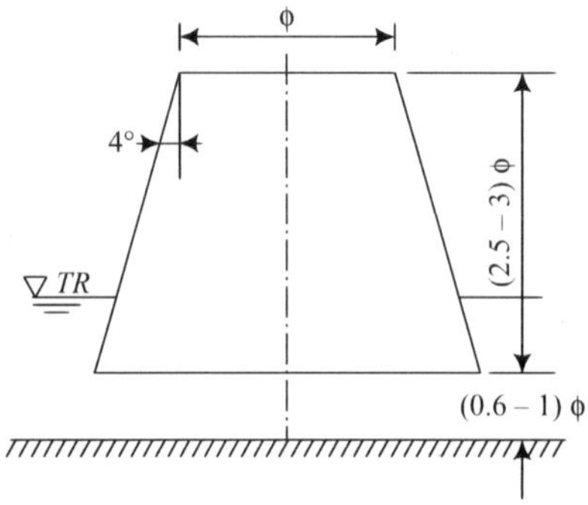

(a) Straight divergent tube

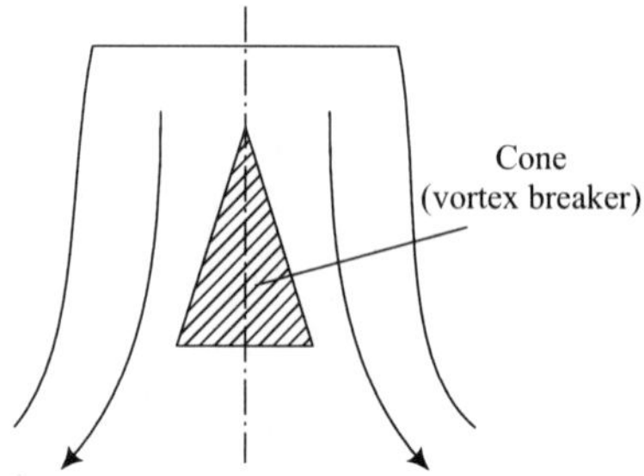

(b) Bell mouth

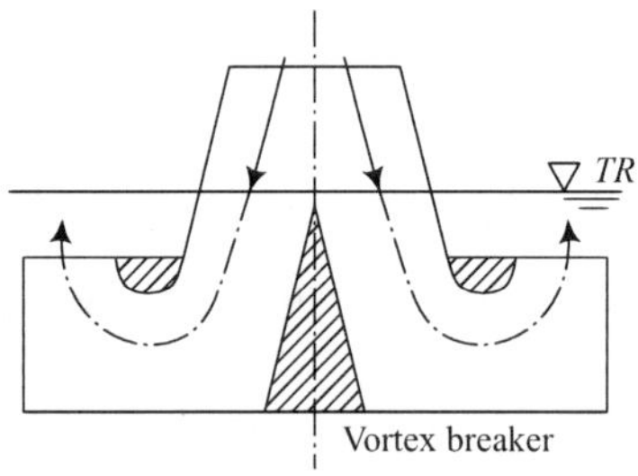

(c) Hydro cone

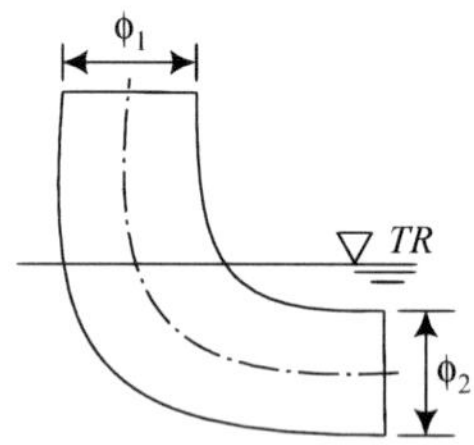

(d) Simple elbow tube

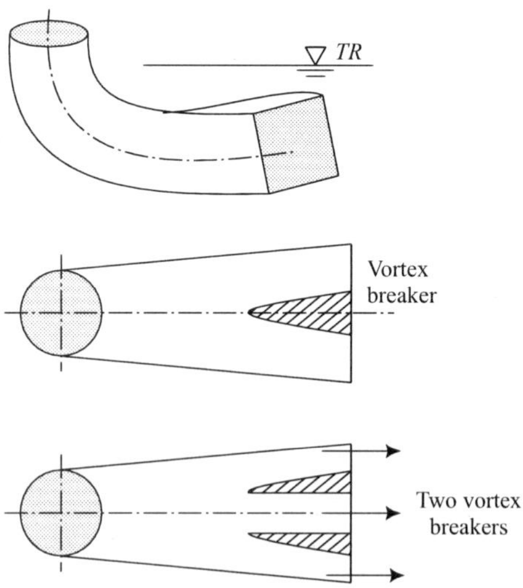

(e) Elbow tube with circular inlet and rectangular outlet

Fig. 6.5 Types of draft tubes

to temperature, water boils resulting in formation of vapours or small bubbles. When these bubbles are carried away by water to a high-pressure zone, these condense and bubbles collapse suddenly. There is a cavity (discontinuity) in the flow and surrounding liquid rushes towards cavity and collides to give very high local pressure (>7000 bar). The process is repeated several times in a second. There can be two effects.

1. Metallic pitting

The vacuum cavities if formed on a metal surface of turbine parts (mainly runner and draft tube), can cause pitting on metal surface. This process is called cavitation of metal surface. The effect can be very serious resulting in pitting surface, fatigue failure and tearing of metal parts if not attended in time. The roughened surface affects the streamline flow pattern leading to high friction losses and lower turbine efficiency.

2. Vibration

If vacuum cavities are formed away from metal surface, pressure waves are formed. This results in water hammer, vibrations and noise.

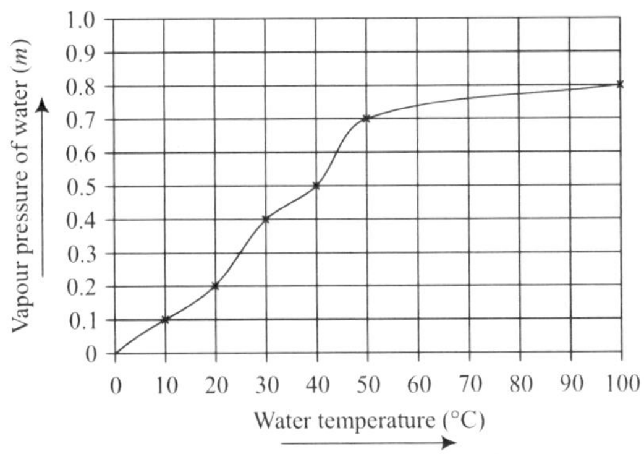

Fig. 6.6 Vapour pressure of water

6.5 FACTORS CAUSING CAVITATION

The turbine should be properly designed to avoid cavitation as it has very serious consequences. The turbine life is affected and efficiency is lowered.

The cavitation depends upon the following factors:

1. Vapour pressure and temperature of water.
2. Mean sea level of turbine installation and atmospheric pressure (H_a).
3. Pressure and velocity of water at the runner exit.

The cavitation in reaction turbine is decided by specific speed (N_S) and sigma factor (σ).

$$\sigma = \frac{(H_a - H_v) - H_s}{H}$$

where H_a = Atmospheric pressure function of attitude.

H_V = Vapour pressure function of water temperature

H_S = Height of runner outlet above tail race

H = Working head of turbine.

Cavitation can be avoided if σ is more than σ_{critical}.

As per Prof. P.H. Roger of USA

$$\sigma_{\text{critical}} = 0.0317\left(\frac{N_s}{100}\right)^2$$

or

$$N_S = 100\sqrt{\frac{\sigma_{critical}}{0.0317}}$$

The critical cavitation condition is expressed by Thomas parameter which is defined as $\frac{NPSH}{H}$.

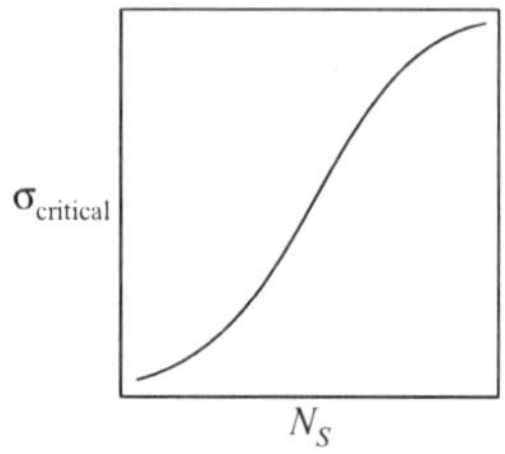

Fig. 6.7 σ_{critical} v/s N_S

6.6 METHODS TO AVOID CAVITATION

The following methods can be used to avoid cavitation

1. Installation of turbine below tail race level

The value of σ can be increased as static head H_S decreases (or even becomes negative). However, maintenance of turbine becomes complicated as water has to be pumped out for inspection and repair.

2. Cavitation free runner

The turbine models and blades are subjected to testing simulating actual working conditions. The points of low pressure are found out and design modifications are made to avoid critical low pressure areas.

3. Use of suitable materials

(i) Cast iron has less resistance to cavitation. It is also difficult to repair the damage by welding.

(ii) Cast steel has more resistance to cavitation. It can also be repaired by welding.

(iii) Stainless steel (Ni-Cr) has maximum resistance to corrosion and cavitation. However, it is very expensive. The turbine runners are made of cast steel and overlaid with stainless steel as cavitation is a surface phenomenon.

(iv) For chemically active water, bronze is a better material which has good cavitation resistance. It can also be repaired by brazing.

(v) Steel plates have better resistance to cavitation than steel castings.

4. Machining of surface

Turbine blades are finely polished to improve cavitation resistance. Rough surface can cause interruption to stream flow lines leading to vortices and hence cavitation.

QUESTION BANK NO. 6

1. What are the uses of draft tube? Describe with neat sketches different types of draft tubes.
2. On what factors does the cavitation in water turbines depend?
3. Describe some methods to avoid cavitation in water turbines.

TUTORIAL SHEET NO. 6

1. A Francis turbine is installed at a site where atmospheric pressure is 10 m of H_2O and vapour pressure is 0.20 m of H_2O. The turbine is designed to produce 11760 kW under a head of 25 m when running at 120 rpm. Calculate the maximum height of straight draft tube for the turbine.

Solution Specific speed,

$$N_S = \frac{N\sqrt{P_t}}{H^{5/4}}$$

$$= \frac{120\sqrt{11760}}{25^{5/4}} = 232.8 \text{ rpm}$$

Critical value of Thomas's cavitation factor for Francis turbine,

$$\sigma_{\text{crit}} = 0.044\left(\frac{N_s}{100}\right)^2 = 0.044\left(\frac{232.8}{100}\right)^2 = 0.238$$

The cavitation factor,

$$\sigma = \frac{(H_a - H_v) - H_s}{H} = \frac{(10 - 0.20) - H_s}{25}$$

$$\sigma > \sigma_{\text{crit}} \text{ to avoid cavitation,}$$

If $\sigma = \sigma_{\text{crit}}$

$$0.238 = \frac{(10 - 0.20) - H_s}{25}$$

∴ $H_S = 3.85$ m.

Maximum permissible height of draft tube.

2. A Kaplan turbine is developing 3250 kW under a head of 6 m. A conical draft tube with inlet diameter of 2.8 m is placed 1.5 m above tail race water and has an efficiency of 76%. The pressure at inlet to draft tube is 5 m H_2O and atmospheric pressure of 10.3 m H_2O. Calculate the efficiency of turbine.

Solution

$$\frac{p_1}{\rho g} = \frac{p_a}{\rho g} - H_S\left(\frac{V_1^2 - V_2^2}{2g} - h_f\right)$$

Neglect $h_f = 0$

$$(10.3 - 5) = 10.3 - 1.5\left(\frac{V_1^2 - V_2^2}{2g} - 0\right)$$

∴ $$\frac{V_1^2 - V_2^2}{2g} = 3.5$$

The efficiency of draft tube,

$$\eta_d = \frac{(V_1^2 - V_2^2)/2g}{V_1^2/2g}$$

∴ $$V_1^2 = \frac{(V_1^2 - V_2^2)}{\eta d} = \frac{3.5}{0.76}$$

$$V_1 = 9.50 \text{ m/s.}$$

The water flow rate,

$$Q = \frac{\pi}{4} D_1^2 \times V_1 = \frac{\pi}{4}(2.8)^2 \times 9.50 = 58.47 \text{ m}^3/\text{s.}$$

∴ The turbine power

$$P_t = \rho Q g H \eta_t$$

∴
$$\eta_t = \frac{P_t}{\rho Q g H} = \frac{3250 \times 10^3}{1000 \times 9.81 \times 58.47 \times 6}$$

$$= 0.945 = 94.5\%$$

CHAPTER 7

Governing of Hydraulic Turbines

7.1 INTRODUCTION

Hydraulic turbines are directly coupled to the electric generators which must run at constant designed speed to generate power at required constant frequency. The constant speed of the generator is called the synchronous speed and is given by,

$$N = \frac{60f}{p} \text{ (rpm)}$$

where f = frequency of the power generated (cps)

p = number of pair of poles of the generator

For Indian conditions f is 50 cps and frequency variation allowed may be $\pm$ 0.15, i.e., 49.85 to 50.15 cps.

Therefore, speed governors are provided on the turbine which control the rate of flow of water into the turbine corresponding to the load in order to keep the speed constant.

The regulation system should be quick acting and accurate. On the other hand, it should be stable without too much sensitivity causing hunting.

The above conflicting requirements of speed regulation system are met by a feedback system as shown.

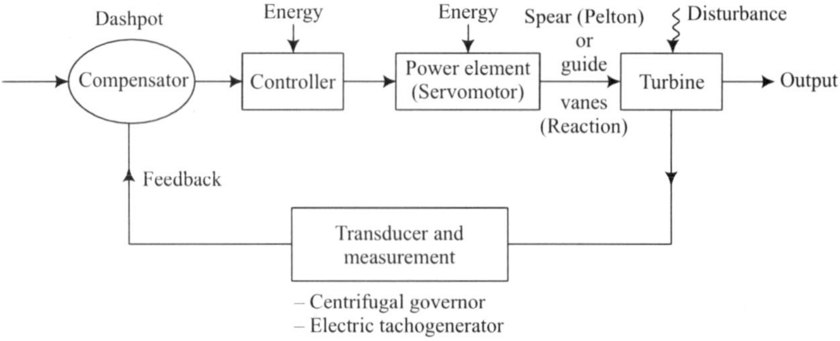

Fig. 7.1 Feedback speed regulation system

7.2 OIL PRESSURE GOVERNOR

The schematic diagram of a simple oil pressure speed governor is shown in Fig. 7.2. The main elements are:

1. Actuator or pendulum
2. Relay valve
3. Servomotor

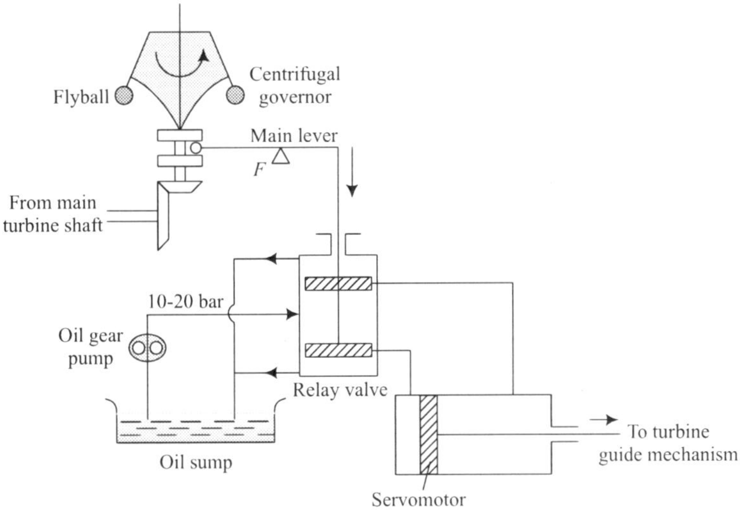

Fig. 7.2 Oil pressure governor

1. Actuator or pendulum

The flyballs of centrifugal governor respond quickly to speed variation of turbine shaft and actuates the main lever.

Modern turbines use electric tachogenerator which is extremely sensitive to frequency variation.

2. Relay valve

It is a piston type slide valve. The motion of the flyballs is transmitted to relay valve which distributes oil to either side of servomotor.

3. Servomotor

Servomotor is a piston cylinder arrangement which transmits power to turbine guide mechanism. In case of Pelton turbine, the guide mechanism is the spear rod. In case of Francis turbine and Propeller turbine, the guide mechanism is guide vanes. In case of Kaplan turbine, the power is transmitted to both guide vanes as well as runner vanes.

7.3 DOUBLE REGULATION SYSTEM OF PELTON TURBINE

The water flow to the runner of a Pelton turbine is regulated by the combined action of the spear and the deflector plate as shown in Fig. 7.3.

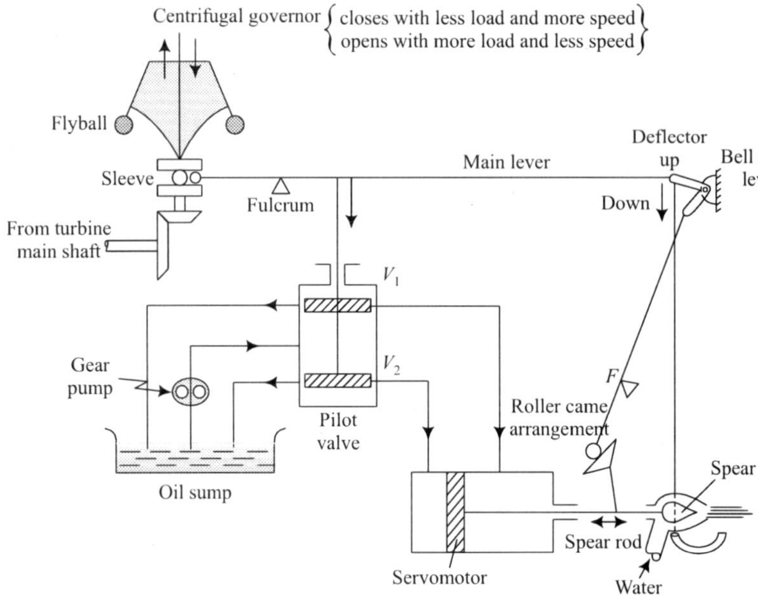

Fig. 7.3 Double regulation system of Pelton turbines

The sensitivity of centrifugal governor is augmented by a hydraulic servomotor.

(a) When the load on the generator drops, the speed of the turbine increases. The flyballs of the centrifugal governor fly outward due to more centrifugal force (due to higher speed). The sleeve moves up pushing the piston rod of pilot valve downwards. The valve V_1 closes and valve V_2 opens. A gear pump supplies oil from an oil sump to the pilot valve. The oil flows to the left chamber of servomotor. The spear rod moves to the right, thereby decreasing the flow area of nozzle and the rate of water flow to the turbine is reduced. The turbine speed falls to normal value and flyballs, sleeve lever, etc. come to normal position.

(b) When the load on the generator increases, speed decreases, flyballs move inward, the sleeve moves down, the piston rod of pilot valve moves up, valve V_2 closes, valve V_1 opens, pressure oil is supplied to right chamber of servomotor the spear rod moves to left, the nozzle opens more, more water flows to the turbine to take up increased load, speed decreases to its rated value.

(c) When there is sudden fall of load, the spear moves rapidly to close the nozzle. There can be water hammer in the penstock due to sudden stoppage of water flow. In order to avoid sudden stoppage of water flow and water hammer, the spear is designed to allow some water flow during sudden load fall. The bell crank lever moves down and deflector moves up. The water is deflected and not allowed to strike the buckets. The deflected water goes as waste to the tail race.

7.4 GOVERNING OF REACTION TURBINES

The guide blades are pivoted and connected to regulating ring which in turn is connected to the piston rod of servomotor of oil pressure governor through regulating rods and regulating lever. When the servomotor moves to the right, the guide blades are rotated to decrease or close the water supply to the runner. The reverse happens when the servomotor moves to the right under decrease of generator load and increase of turbine speed. The guide blades are rotated to increase the water flow area to the turbine runner.

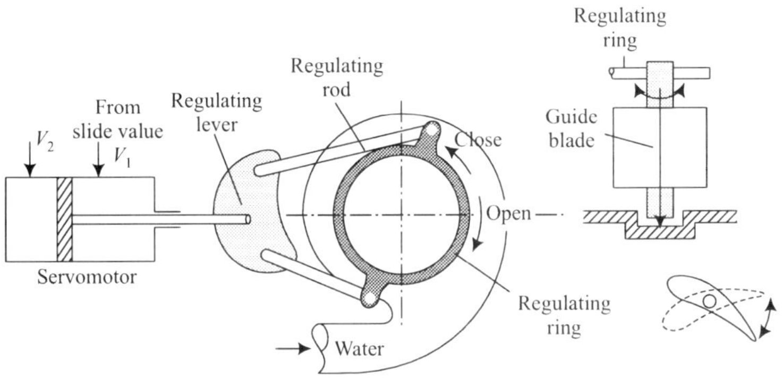

Fig. 7.4 Governing of reaction turbine

In case of Kaplan turbine both guide vanes and runner blades are adjusted simultaneously in case of load fluctuation. There is an additional servomotor and a control valve interconnected to those of guide vanes so that for a given guide vane opening, there is a definite runner blade inclination. The runner blade operating mechanism is housed in the hub of the runner which is hollow.

When the load suddenly falls on reaction turbine, a pressure relief valve is operated by speed governor. A portion of water from spiral casing is allowed to flow directly through relief valve to the tail race without striking the turbine runner. The pressure relief valve performs the function of protection system to avoid water hammer effects during sudden load drop.

QUESTION BANK NO. 7

1. What do you understand by governing of hydraulic turbine? Explain with sketches, the working of an oil pressure governor.
2. Describe, in brief, with the help of neat sketch, method of governing of Pelton wheel. How is water hammer avoided in the penstock during sudden load throw off?
3. Explain with the help of neat sketches, method of governing of Francis turbine. How is it different in case of Kaplan turbine?
4. Compare the protection system against water hammer due to sudden load drop in case of impulse turbine and reaction turbine.
5. What is feedback speed governing system? What are the main requirements of governing system?

CHAPTER 8

Dynamic Similarity and Performance Characteristics

8.1 INTRODUCTION

The type, size and number of hydro turbines for a power plant are selected on the basis of detailed techno-economical analysis of site conditions. The performance of a prototype (first of the series of turbine) cannot be predicted before its manufacture without its testing in the laboratory or on a test bed. It is very difficult to carry out its tests because of its size and facilities required such as variable head, speed and load under which the turbine is expected to operate at the power plant. It is very costly to affect changes on a full size machine and testing can be time-consuming.

It is economical to make scale models and perform necessary hydraulic tests on it in order to predict the performance of prototype or full-sized machine under similar operating conditions.

The models are prepared and tested within the following ranges:

The model capacity, $P_m = 5$ to 50 kW.

The runner diameter, $D_m = 250 - 600$ mm.

Test head for model, $H_m = 4 - 100$ m.

8.2 DYNAMIC SIMILARITY OF MODEL AND PROTOTYPE

The turbine model should satisfy the following similarities with the prototype.

1. Geometric similarity

There should be constant linear proportions and same angles between model and prototype.

$$(\alpha_1, \beta_1, \alpha_2, \beta_2)_m = (\alpha_1, \beta_1, \alpha_2, \beta_2)_p$$

$$\left(\frac{D_1}{D_2}\right)_m = \left(\frac{D_1}{D_2}\right)_p$$

where m = model

p = prototype.

2. Kinematic similarity

There should be same flow patterns and same velocity ratios.

$$\phi_m = \phi_p$$
$$\psi_m = \psi_p$$
$$\xi_m = \xi_p$$

The velocity triangles of model and prototype should be congruent.

3. Dynamic similarity

The Reynolds number

$$\text{Re}_m = \text{Re}_p$$

There should be proportionality of forces acting on the runner blades.

If the above conditions are fulfilled, the following parameters are achieved and $\eta_m = \eta_p$.

Model	*Prototype*
1. $Q_m = \frac{\pi}{4} d_{jm}^2 V_{jm}$	1. $Q = \frac{\pi}{4} d_j^2 V_j$
2. $P_m = \rho Q_m g H_m \eta_m$	2. $P = \rho Q g H \eta$
3. $V_{bm} = \frac{\pi D_m N_m}{60}$	3. $V_b = \frac{\pi DN}{60}$

8.3 SPECIFIC SPEED

The suitability of a turbine for a given application depends on:

1. Head of water, H (m)
2. Rotational speed, N (rpm)
3. Power developed, P_t (kW)

The above parameters constitute a characteristic called *specific speed.*

The specific speed of a turbine is defined as the speed of operation of a *geometrically similar model* of the turbine which is so proportioned that it produces 1 kW power when operating under a head of 1 m.

The blade speed,

$$V_b = \frac{\pi DN}{60}$$

$$\frac{V_b}{V_j} = \phi$$

$\therefore \quad V_b = \phi V_j = \phi\sqrt{2gH}$

$\therefore \quad V_b \propto \sqrt{H}$

But $\quad V_b \propto DN$

$\therefore \quad DN \propto \sqrt{H}$ ...(1)

The turbine power,

$$P_t = \rho QgH$$

The density of power ρ and acceleration due to gravity g are constant.

$\therefore \quad P_t \propto QH$

The water discharge,

$$Q = \frac{\pi}{4}D^2 V_f$$

$$V_f = \psi\sqrt{2gH}$$

$\therefore \quad V_f \propto \sqrt{H}$

$\therefore \quad Q \propto D^2\sqrt{H}$ ...(2)

$\therefore \quad P_t \propto D^2\sqrt{H} \ldots H$ ...(3)

From equation (3),

$$D^2 \propto \frac{P_t}{H^{3/2}}$$

or $\quad D \propto \dfrac{\sqrt{P_t}}{H^{3/4}}$

From equation (1),

$$N \propto \frac{\sqrt{H}}{D}$$

$$\propto \frac{H^{1/2} \cdot H^{3/4}}{\sqrt{P_t}}$$

$$\propto \frac{H^{5/4}}{\sqrt{P_t}}$$

$$= K\frac{H^{5/4}}{\sqrt{P_t}}$$

where K = Proportionality constant

$$= \frac{N\sqrt{P_t}}{H^{5/4}}$$

If $$P_t = 1 \text{ kW}, \; H = 1 \text{ m},$$

$$K = N = N_S$$

where N_S is called specific speed.

$$\therefore \quad N_S = \frac{N\sqrt{P_t}}{H^{5/4}}$$

For equal efficiencies of model and prototype,

$$(N_S)_m = (N_S)_p.$$

The classification of turbines on the basis of specific speed has been discussed in the previous chapter.

8.4 SCALE RATIO

The model of a turbine and its prototype should be in a definite geometric ratio depending upon their respective heads and rotational speeds. The ratio of blade velocity V_b and water velocity $V(V_j$ in Pelton wheel and V_a in reaction turbines) is called speed ratio. Different types of turbines have definite value of speed ratio.

Pelton turbine, $\phi = 0.42$ to 0.47

Francis turbine, $\phi = 0.55$ to 1.00

Kaplan turbine, $\phi = 1.50$ to 3.00

From equation (1) of # 8.3,

$$DN \propto \sqrt{H}$$

$$\therefore \quad \frac{D_m N_m}{D_p N_p} = \sqrt{\frac{H_m}{H_p}}$$

$$\frac{D_m}{D_p} = \sqrt{\frac{H_m}{H_p}} \cdot \frac{N_p}{N_m} \qquad \ldots(4)$$

The scale ratio $\left(\frac{D_m}{D_p}\right)$ is the ratio of diameter of model of turbine and prototype of turbine.

8.5 UNIT QUANTITIES

The operational characteristics of hydraulic turbines are expressed in terms of unit quantities. These quantities help to study the performance of turbine independent of actual head, discharge and power output.

1. Unit speed

Unit speed (N_U) is defined as the speed of a geometrically similar turbine (model) working under a head of 1 m.

$$V_b = \phi\sqrt{2gH}$$

$$\frac{\pi DN}{60} = \phi\sqrt{2gH}$$

$$DN \propto \sqrt{2gH}$$

$$N \propto \sqrt{H}$$

$\therefore$ $$N = K\sqrt{H}$$

where K = constant

when $H = 1$ m, $N = N_U = K$

$\therefore$ Unit speed,

$$N_U = \frac{N}{\sqrt{H}} \qquad \text{...(5)}$$

2. Unit power

The unit power (P_U) is the power in kW generated by a geometrically similar turbine (model) under a head of 1 m,

$$P_t = \rho QgH$$

$$= \rho A\sqrt{2gH} \cdot gH$$

ρ and g are constant.

$\therefore$ $$P_t \propto H^{3/2}$$

$\therefore$ $$P_t = KH^{3/2}$$

when $H = 1$ m, $$P_t = P_U = K$$

$\therefore$ $$P_U = \frac{P_t}{H^{3/2}} \qquad \text{...(6)}$$

3. Unit discharge

The unit discharge (Q_U) is the flow rate of a geometrically similar turbine (model) under a head of 1 m.

$$Q = A\sqrt{2gH}$$

$$= K\sqrt{H}$$

when $H = 1$ m, $\qquad Q = Q_U = K$

$$\therefore \qquad Q_U = \frac{Q}{\sqrt{H}} \qquad \text{...(7)}$$

8.6 MODEL RELATIONSHIPS

Model relationships are used to predict the behaviour of a prototype from the test runs of model of turbine.

1. Head coefficient

The blade velocity,

$$V_b = \phi\sqrt{2gH} = \frac{\pi DN}{60}$$

$$\therefore \qquad N = \frac{60\phi\sqrt{2gH}}{\pi D}$$

$$\propto \frac{\sqrt{H}}{D}$$

$$\therefore \qquad ND \propto \sqrt{H}$$

$$\therefore \qquad \frac{H}{N^2D^2} = \text{constant.} \qquad \text{...(8)}$$

The parameter $\frac{H}{N^2D^2}$ is called head coefficient.

2. Flow coefficient

The flow rate of water through a turbine

$$Q = A \times V_f$$

$$A \propto D^2$$

$$V_f = \psi\sqrt{2gH}$$

$$\propto \sqrt{H}$$

$$\therefore \qquad Q \propto D^2\sqrt{H}$$

From equation (8),

$$H \propto N^2D^2$$

$\therefore$
$$Q \propto ND^3$$

or
$$\frac{Q}{ND^3} = \text{constant} \qquad ...(9)$$

The parameter $\frac{Q}{ND^3}$ is called capacity or flow coefficient.

3. Power coefficient

The power developed by a turbine,

$$P_t = \rho g Q H \eta_t$$
$$\propto QH$$

But
$$Q \propto ND^3$$

and
$$H \propto N^2D^2$$

$\therefore$
$$P_t \propto ND^3 \times N^2D^2$$

$\therefore$
$$P_t \propto N^3D^5$$

or
$$\frac{P_t}{N^3D^5} = \text{constant.}$$

The parameter $\frac{P_t}{N^3D^5}$ is called power coefficient.

8.7 PERFORMANCE CHARACTERISTICS

Hydraulic turbines are designed to operate at certain values of head, discharge, power, speed and efficiency. However, turbines have to operate at off design conditions also. Therefore, the performance of turbines at design conditions and varying conditions have to be determined by conducting tests on turbine models or prototypes.

1. Parameters and properties

The following parameters and properties are collected from model tests.

(i) Flow rate, $Q(\text{m}^3/\text{s})$ or mass flow rate $\dot{m}_f (= \rho Q)$ in kg/s.

(ii) Turbine power, $P_t = \frac{\rho Q g H \eta_t}{10^3}$ (kW)

(iii) Efficiencies, η_{trans}, η_n, η_{vol}, η_h, η_m and η_t,

(iv) Turbine speed, N(rpm)

The data is collected for design conditions and for varying conditions of operation.

2. Graphical representation

The results of the tests are plotted graphically and the curves are called performance curves or *characteristic curves*. Six points are required to plot a curve and two points for a straight line as per Indian standard code. The representation can be made on:

(i) Cartesian coordinates
(ii) Log scale or Log Log scales
(iii) Circular graphs

3. Measurements

(i) Flow rate by flow meters
(ii) Pressure in m of H_2O or kPa by pressure gauge
(iii) Speed in rpm or rps by speedometer
(iv) Output in kW by dynamometer either absorption type or transmission type.

4. Type of characteristics

For the sake of convenience, the characteristic curves are plotted in unit quantities. Following types of curves are plotted.

(i) *Main characteristics* or constant head characteristic curves.

Tests are performed by maintaining a constant head and a constant nozzle or guide valve opening. Tests are carried at $\frac{1}{4}, \frac{1}{2}, \frac{3}{4}$ and full opening. Similar tests are conducted for each gate opening. A series of values of N, Q and P are measured. The unit quantities of Q_U, P_U, N_U and η_t are computed and plotted as shown in Fig. 8.1. The following types of curves are obtained.

Pelton turbine : Horizontal straight line
Francis turbine : Drooping curve
Kaplan turbine : Parabolic

(ii) *Operating characteristic* or constant speed curves.

Constant speed is obtained by regulating the gate opening (i.e., discharge) as the load varies. The following values are measured and calculated. Q_0 is the discharge required to initiate the motion of turbine runner from state of rest. For constant head $P_U \propto Q_U$ and a straight line is obtained. However, η_U increases with Q_U and becomes more or less constant beyond a certain value.

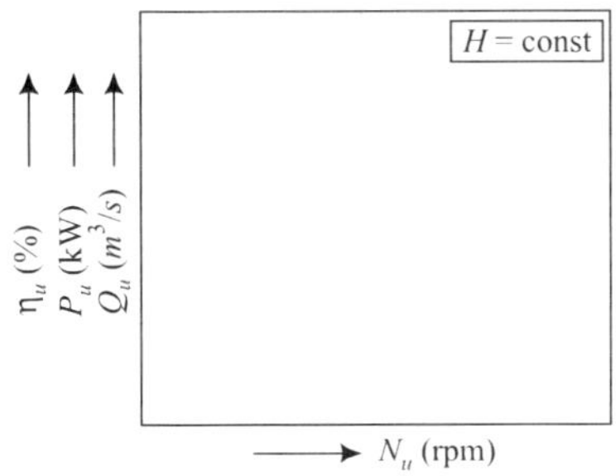

Fig. 8.1 Main characteristics

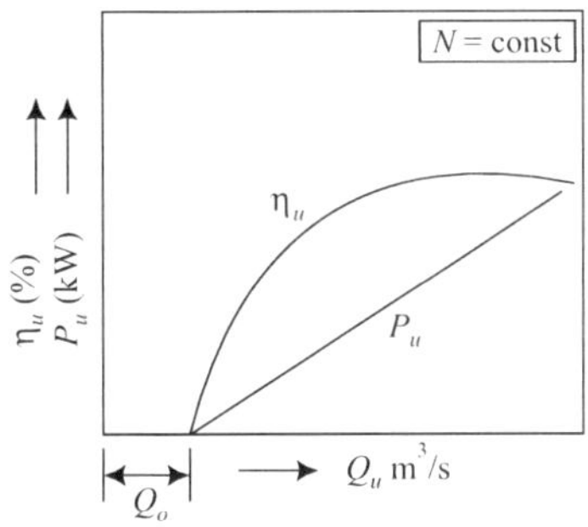

Fig. 8.2 Operating characteristics

(iii) *Universal characteristic* or constant efficiency curves

These curves are plotted between Q_U and N_U for different gate openings. These curves are used for optimum design of turbine. The hydraulic design is for optimum speed but it must satisfy the safety conditions of run away speed.

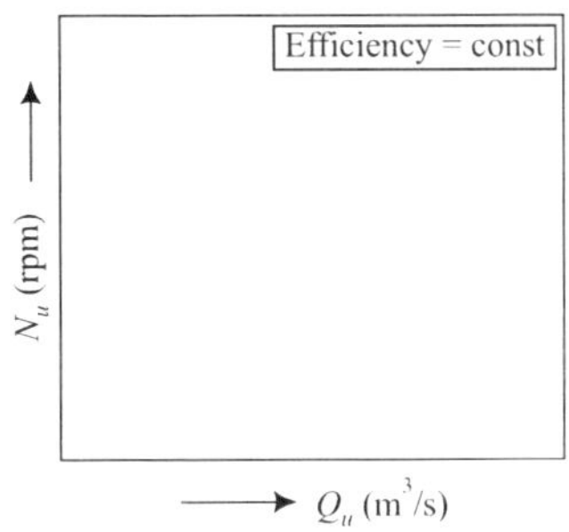

Fig. 8.3 Universal characteristic

5. Appreciation of characteristic curves

Appreciation of characteristic curves is fitting of curves to mathematical equations for feeding to computer programming. The following types of curves are obtained.

(i) Straight line

$$y = mx + c$$

(ii) Rectangular parabola

$$x^2 - y^2 = c$$

(iii) Drooping curve or polynomial

$$y = a_0 + a_1 x + a_2 x^2 + \ldots + a_n x^n$$

(iv) Rising curve

$$y = bx^m$$

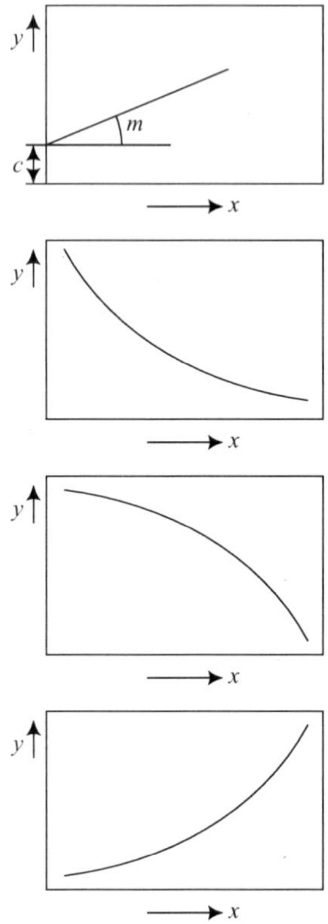

Fig. 8.4 Different curves

6. Selection of turbine

The type of turbine, output and speed are selected for best performance. This has been discussed in the previous chapters.

QUESTION BANK NO. 8

1. What is the need for model testing? Explain dynamic similarity.
2. What are specific speed and unit quantities?
3. Explain scale ratio and model relationships.
4. What are performance characteristics? How are these prepared?
5. Explain the following:
   (i) Main characteristics

(ii) Operating characteristics
(iii) Universal characteristics

TUTORIAL SHEET NO. 8

1. Determine the speed and power of a Francis turbine operating under a head of 30 m if the $\frac{1}{5}$th scale model could develop 5 kW under a head of 1.5 m and 450 rpm.

Solution $$\frac{D_m}{D_p} = \frac{N_p}{N_m}\sqrt{\frac{H_m}{H_p}}$$

$$N_p = \frac{D_m}{D_p}\cdot\sqrt{\frac{H_p}{H_m}}\cdot N_m$$

$$= \frac{1}{5}\times\sqrt{\frac{30}{1.5}}\cdot 450 = \mathbf{402.5\ rpm} \quad \text{Ans}$$

$$N_S = \frac{N_m\sqrt{P_m}}{H_m^{5/4}} = \frac{450\sqrt{5}}{(1.5)^{5/4}} = 606.16$$

Also $$N_S = \frac{N_p\sqrt{P_p}}{H_p^{5/4}}$$

$\therefore$ $$\sqrt{P_p} = \frac{N_S H_p^{5/4}}{N_p} = \frac{606.16\times 30^{5/4}}{402.5} = 105.736$$

$\therefore$ $P_p = \mathbf{11180\ kW}$ Ans.

2. A Francis turbine with specific speed of 210 rpm has to develop 30 MW at 180 rpm. An experimental model has to be prepared to operate at 4.5 m head with a flow rate of 0.6 m^3/s. Assuming an efficiency of 88%, estimate the speed, power and scale ratio for the model. What will be the flow rate for the turbine?

Solution The power developed by the model,

$$P_m = \rho Q_m g N_m \eta_m \times 10^{-3}$$
$$= 1000\times 0.6\times 9.81\times 4.5\times 0.88\times 10^{-3} = 23.3\ \text{kW}$$

$$N_S = \frac{N_m\sqrt{P_m}}{H_m^{5/4}}$$

$$N_m = \frac{N_S \times H_m^{5/4}}{\sqrt{P_m}} = \frac{210 \times 4.5^{5/4}}{\sqrt{23.3}} = 285 \text{ rpm}$$

$$\frac{D_p N_p}{D_m N_m} = \sqrt{\frac{H_p}{H_m}}$$

Scale ratio, $$\frac{D_p}{D_m} = \frac{N_m}{N_p}\sqrt{\frac{H_p}{H_m}}$$

Again, $$N_S = \frac{N_p\sqrt{P_p}}{H_p^{5/4}}$$

$\therefore$ $$H_p^{5/4} = \frac{N_p\sqrt{P_p}}{N_S} = \frac{180\sqrt{30 \times 10^3}}{210}$$

$$= 148.46$$

$\therefore$ $$H_p = 54.6 \text{ m}$$

$\therefore$ $$\frac{D_p}{D_m} = \frac{285}{180}\sqrt{\frac{54.6}{4.5}} = \mathbf{5.5} \text{ Ans}$$

$$P_p = \rho Q_p g H_p \eta_p$$

$\therefore$ $$Q_p = \frac{P_p}{\rho g H_p \eta_p} = \frac{30 \times 10^6}{1000 \times 9.81 \times 54.6 \times 0.88}$$

$$= \mathbf{63.65\ m^3/s} \quad \text{Ans}$$

PART III

PUMPS

CHAPTER 9

Centrifugal Pumps

9.1 INTRODUCTION

The centrifugal pumps are the most widely used pumps in industry and other sectors of economy. The list of applications is enclosed in Table 1.1. The centrifugal pumps have low initial cost, simple construction, high efficiency, low installation and operation costs. A centrifugal pump mainly consists of a rotating impeller mounted on a shaft. It has a number of curved vanes. The stationary element enclosing the impeller is volute casing. The impeller is rotated by an electric motor or an engine. The water or other liquid enters the eye of the impeller and is imparted the rotational energy of impeller as kinetic energy. The liquid from impeller is discharged into volute casing which converts kinetic energy into pressure energy. The liquid is supplied to the pump by a suction pipe and is delivered to the delivery pipe.

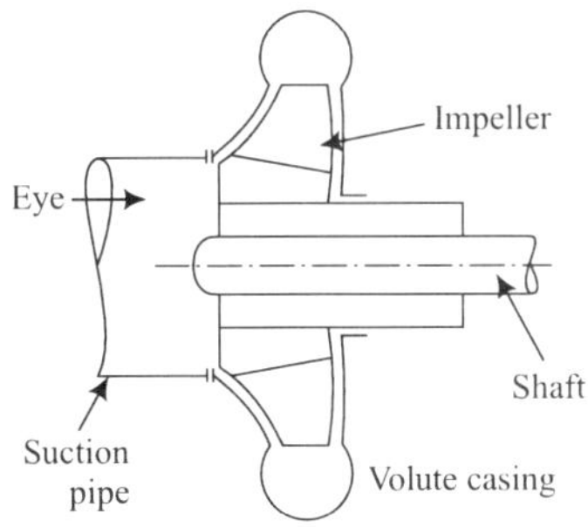

Fig. 9.1 Centrifugal pump

9.2 CLASSIFICATION OF CENTRIFUGAL PUMPS

There is a very large variety of centrifugal pumps. These are classified by the following characteristic features:

1. Commercial classification of working head.
2. Construction
   (i) Type of casing

(ii) Number of impellers
(iii) Flow direction through impeller.
(iv) Number of suction or impeller eyes.
(v) Shaft layout
(vi) Liquid handled and type of impeller

3. Hydraulic classification by specific speed.

9.2.1 Working Head

The centrifugal pumps are commercially classified by the working head or delivery head of liquid

(i) Low head pumps upto 15 m
(ii) Medium head pumps: 15 m to 40 m
(iii) High head pumps: above 40 m.

9.2.2 Shape and Type of Casing

The high kinetic energy of fluid leaving the impeller of the pump is recovered and converted into pressure energy by the casing around the impeller. There are two types of casings.

1. Volute casing or scroll casing

The kinetic energy of liquid is converted into pressure energy by the diffuser action of the gradually increasing area of volute or scroll casing. The pump is called volute pump.

2. Casing with stationary guide vanes

Fixed guide blades are provided around the impeller periphery. The kinetic energy of fluid is converted into pressure energy by the diverging passages formed between the guide vanes. The pump is called *turbine pump* or *diffuser pump*. The turbine pumps are more efficient and used in multi-stage pumps.

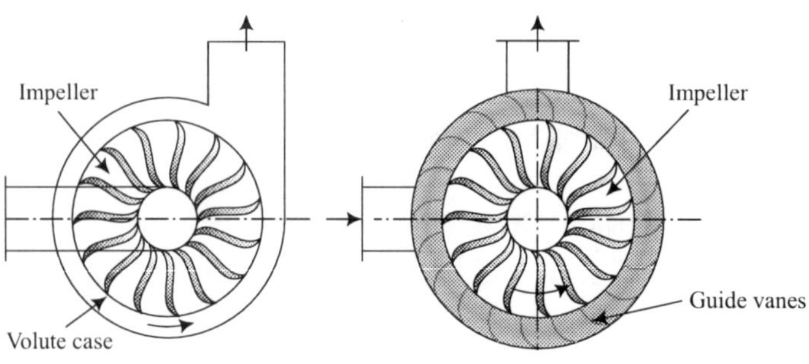

(a) Volute pump (b) Turbine pump (Diffuser pump)

Fig. 9.2 Type of casings

9.2.3 Number of Impellers (Stages)

Single impeller pumps are widely used. But there is a limit on the head developed (15m) by single impeller. For high pressure requirements, the impellers are arranged in series. Such pumps are called multi-stage pumps.

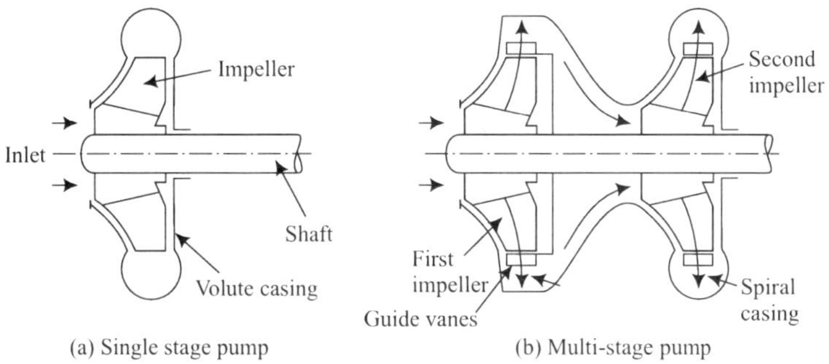

Fig. 9.3 Number of impellers

9.2.4 Flow Direction through Impeller

The liquid enters the impeller axially and head is developed by centrifugal force. The liquid leaves the impeller radially. Such pumps are called radial flow pumps.

In axial flow pumps, the head is developed by the propelling action of vanes. The liquid enters the impeller axially and also leaves axially. These pumps have very large discharge but low head.

Flow through a mixed flow impeller is a combination of axial and radial flow. The head is developed by the combined action of centrifugal force and propelling action of impeller.

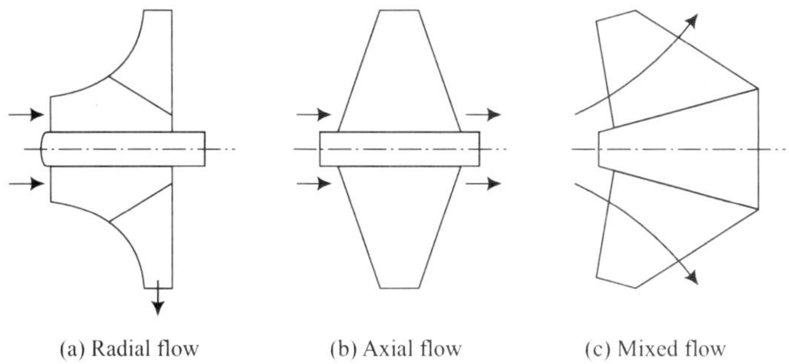

Fig. 9.4 Flow direction through impeller

9.2.5 Number of Inlets

The liquid enters the impeller with single suction. There are pumps with double suction and liquid enters the impeller from both sides. The double suction pumps can handle large discharge and have no axial thrust.

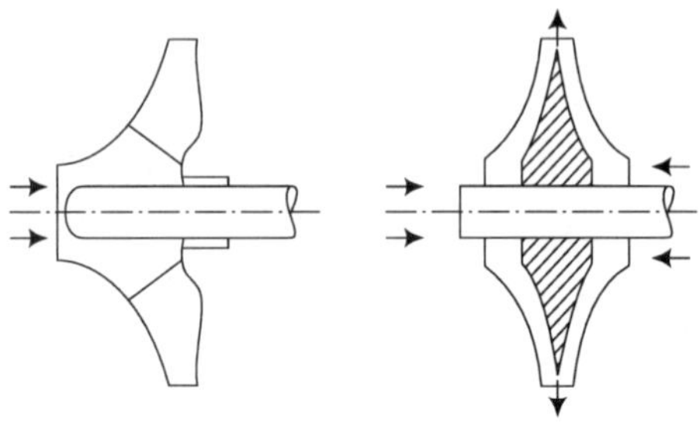

(a) Single section (b) Double section

Fig. 9.5 Single and double suction pumps

9.2.6 Shaft Layout

Most of the centrifugal pumps have horizontal shaft. But to achieve economy of space, vertical shaft pumps are also used. Such pumps are used for deep well and mine pumps.

9.2.7 Liquid Handled

There are three types of impellers:

1. Closed impeller

The vanes are covered with shrouds (side plates) on both sides to provide a smooth passage for the liquid. The design ensures full capacity operation, high efficiency and long life. This type of impeller is suitable for clear liquids of low viscosity, i.e., water, hot water, acids and other clear chemicals.

2. Semi-open impeller

The impeller has shroud on the back side only. It is a non-clog impeller and can handle high viscosity liquids containing fibrous materials such as paper pulp, sugar molasses, sewage water, etc.

3. Open impeller

The impeller has no shrouds and vanes are open on both sides. These pumps are used for very rough duty to handle abrasive liquids, i.e., coal slurry, mixture of water and sand, pebbles and clay. The closed and semi-open impellers are not suitable due to clogging tendency.

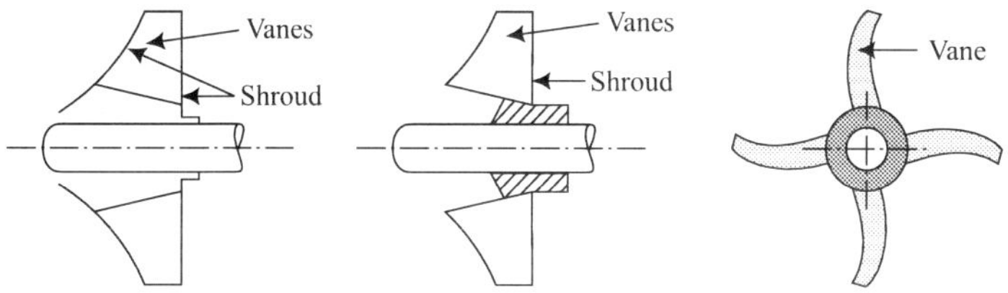

(a) Closed impeller (b) Semi-open impeller (c) Open impeller

Fig. 9.6 Types of impellers

9.2.8 Specific Speed

Specific speed is the speed of rotation of a geometrically similar model of a pump which delivers unit discharge against a unit head.

The specific speed of a pump,

$$N_S = \frac{N\sqrt{Q}}{H^{3/4}} \text{ (rpm)}$$

where N = Pump speed (rpm)

Q = Discharge (m^3/s)

H = Head per stage (m)

The pumps are classified as per specific speed as given in Table 9.1.

Table 9.1 Classification of centrifugal pumps

Pump Type	*Speed*	*Specific speed (rpm)*
Radial Flow	Slow	10-30
	Normal	30-50
	High	50-80
Mixed Flow	–	80-160
Axial Flow	–	160-500

9.3 PUMP PERFORMANCE

The installation of a centrifugal pump is shown in Fig. 9.7 together with valves, piping, fittings and instrumentation. The following nomenclature is used.

1. Suction lift (H_s)

The vertical height between liquid surface and pump impeller centre line.

2. Discharge lift (H_d)

The vertical height between pump impeller centre line and top surface liquid in discharge tank.

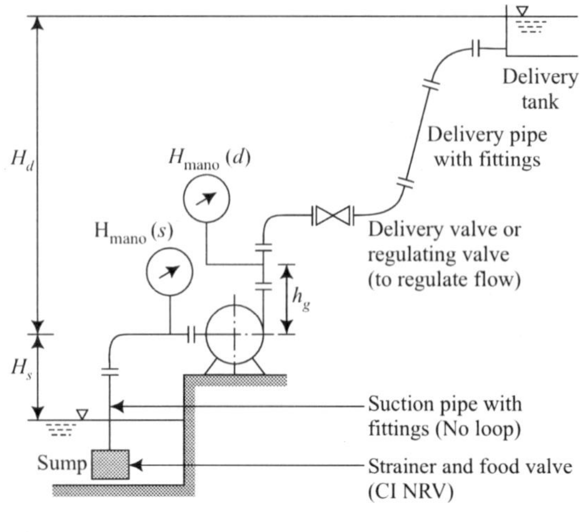

Fig. 9.7 Pump installation

3. Static head (H_{st})

The static head or vertical lift or geometric head represents the total vertical lift.

$$H_{st} = H_s + H_d$$

4. Manometric head (H_{mano})

It is the difference between delivery and suction pressure as measured by a manometer or pressure gauges.

$$H_{mano} = H_{mano(d)} - H_{mano(s)} + h_g$$

5. Total head (H)

Total head is the increase in total energy of liquid between inlet and outlet of the pump.

$$H = H_{mano} + \frac{V_d^2 - V_s^2}{2g}$$

where $$H_{mano} = H_{static} + \Sigma \Delta H_l$$

$$\Sigma \Delta H_l = \text{Total loss of head}$$

6. Pump power (P_i)

The power input required to drive the pump

$$P_i = \frac{\rho Q g H_{mano}}{\eta_{overall}}$$

The overall efficiency of pump installation

$$\eta_{\text{overall}} = \frac{\text{Output}}{\text{Input}} = \frac{\rho Q g H_{\text{mano}}}{\sqrt{3} E I \cos \phi}$$

where I = Electric current (A)

E = Electric voltage (V)

ϕ = Power factor

$$\eta_{\text{static}} = \frac{\rho Q g H_{static}}{P_i}$$

9.4 SANKEY DIAGRAM

The energy flow of the pump installation is shown as Sankey diagram in Fig. 9.8.

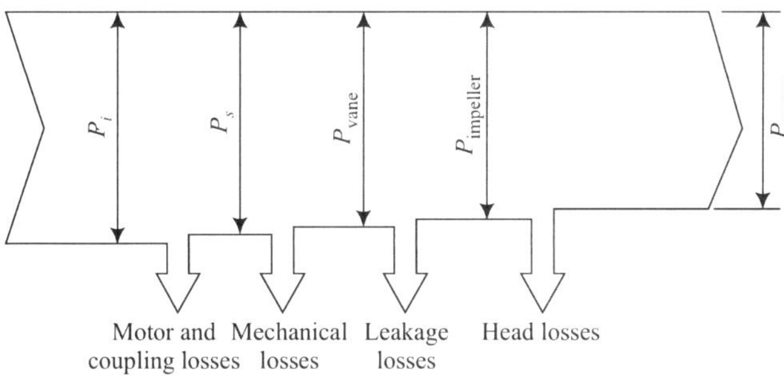

Fig. 9.8 Sankey diagram of pump installation

9.4.1 Losses

The various losses include:

1. Motor and coupling losses

These losses include electrical and mechanical losses in the motor and windage losses of coupling.

2. Mechanical losses

The mechanical losses include:

(i) Disc friction loss of impeller and liquid

(ii) Bearing and gland losses.

3. Leakage losses

Liquid leakage losses in the glands.

4. Head losses

The head losses or hydraulic losses are due to

(i) secondary flow or eddies in the suction eye and impeller outlet
(ii) friction of volute and impeller
(iii) turbulence of liquid flow

9.4.2 Pump Efficiency

1. The overall efficiency or gross efficiency or actual efficiency of the pump set,

$$\eta_{\text{overall}} = \frac{P_{\text{water}}}{P_i} = \frac{\rho Q g H_{\text{mano}}}{\sqrt{3} EI \cos\phi}$$

2. The overall pump efficiency,

$$\eta_{\text{pump}} = \frac{P_{\text{water}}}{P_s}$$

where P_S = Power input to pump shaft
$= P_i -$ Motor and coupling losses

3. The power of the impeller,

$$P_{\text{impeller}} = \frac{V_{w_2} V_{b_2}}{g} \quad \text{(per kg of liquid flow)}$$

4. Volumetric efficiency

$$\eta_{\text{vol}} = \frac{P_{\text{impeller}}}{P_{\text{vane}}} = \frac{Q}{Q + \Delta Q}$$

where Q = Pump discharge
ΔQ = Leakage losses

5. Mechanical efficiency of pump

$$\eta_{\text{mech}} = \frac{\rho(Q + \Delta Q) V_{w_2} V_{b_2}}{P_s} = \frac{P_{\text{vane}}}{P_s}$$

6. Monometric efficiency,

$$\eta_{\text{mono}} = \frac{P_{\text{water}}}{P_{\text{impeller}}} = \frac{\rho Q g H_{\text{mano}}}{\rho Q \left(\dfrac{V_{w_2} V_{b_2}}{g} \right)} = \frac{H_{\text{mano}}}{V_{w_2} V_{b_2}}$$

$$H_{\text{mano}} = H_S + H_f + \frac{V_d^2}{2g}$$

where H_S = Static lift (m)

H_f = Friction losses (m)

$\frac{V_d^2}{2g}$ = Velocity head (m)

7. Hydraulic efficiency

$$\eta_{\text{hyd}} = \frac{\rho Q g H_s}{\rho Q\left(\frac{V_{w_2} V_{b_2}}{g}\right)} = \frac{H_s}{V_{w_2} V_{b_2}}$$

8. The overall efficiency of pump,

$$\eta_{\text{overall}} = \eta_m \times \eta_{\text{vol}} \times \eta_{\text{mano}}$$

9.5 VELOCITY DIAGRAMS

The velocity diagrams at the inlet and outlet of the impeller of a centrifugal pump are shown in Fig. 9.9. The following nomenclature is used:

V_a = Absolute velocity of liquid, (m/s)

V_b = Peripheral velocity of vane, (m/s)

V_r = Relative liquid velocity (m/s)

V_f = Flow component of absolute velocity (m/s)

V_w = Whirl or tangential component of absolute velocity (m/s)

1. suffix for inlet of impeller
2. suffix for outlet of impeller

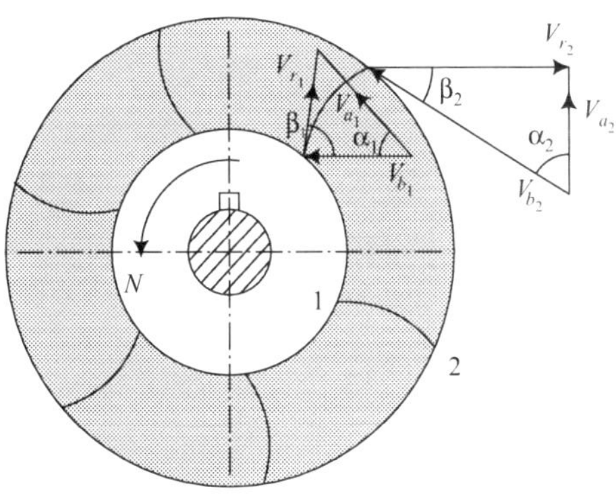

Fig. 9.9 Velocity diagrams of centrifugal pump

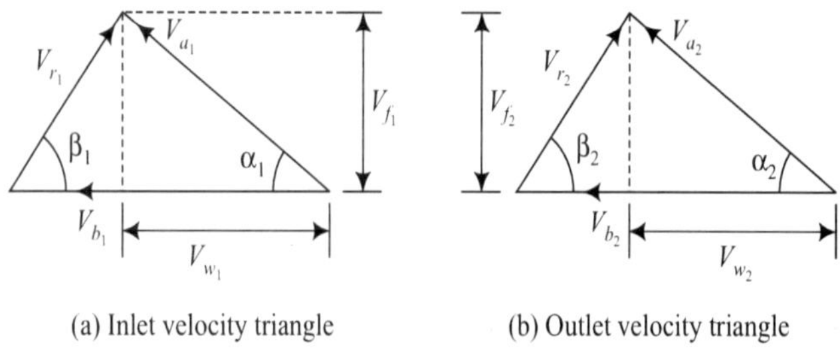

Fig. 9.9 Inlet and outlet velocity triangles

The power applied by impeller on liquid,

$$P_{\text{impeller}} = \rho Q(V_{w_2}V_{b_2} - V_{w_1}V_{b_1})$$

For centrifugal pumps, liquid enters radially into the impeller

$$\therefore \qquad \alpha_1 = 90°$$

$$P_{\text{impeller}} = \rho Q V_{w_2} V_{b_2}.$$

9.5.1 Practical Design Data

For centrifugal pumps, the following data holds good.

$\alpha_1 = 90°$; $\beta_1 = 15°$ to $30°$; $\beta_2 = 20°$ to $40°$

$V_{f_1} = V_{f_2}$; $B \times D =$ constant

No. of vanes: 6 to 12

$$V_{b_2} = \sqrt{2gH} = \frac{\pi D_2 N}{60}$$

$$\therefore \qquad D_2 = \frac{60}{\pi N} \cdot \sqrt{2gH}$$

The inlet diameter of impeller, D_1 depends upon the pump head.

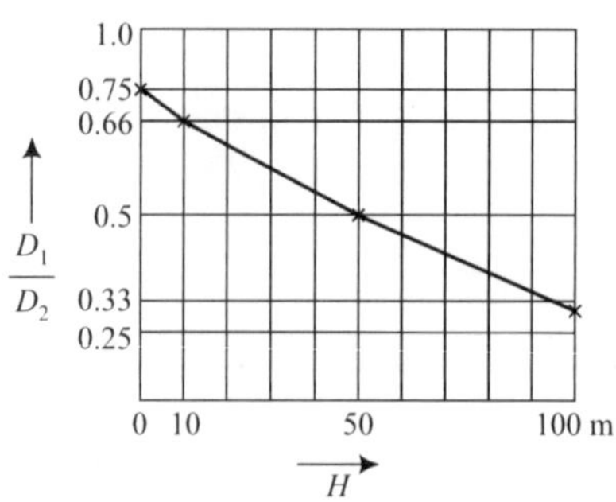

Fig. 9.10 D_1/D_2 v/s H

9.5.2 Effect of Exit Blade Angle β_2

The shape of impeller vane changes with β_2 which affects the efficiency of the pump.

For $\quad \alpha_1 = 90°,$

$$P_{\text{impeller}} = \rho Q (V_{w_2} V_{b_2})$$

$$= \rho Q g H_{\text{mon}}. \quad \text{(If no internal losses)} \qquad ...(1)$$

From outlet velocity triangles, Fig. 9.9(b),

$$V_{w_2} = V_{b_2} - \frac{V_{f_2}}{\tan \beta_2}$$

But $\quad Q = \pi D_2 B_2 V_{f_2}$

$\therefore \quad V_{f_2} = \dfrac{Q}{\pi D_2 B_2}$

$\therefore \quad V_{w_2} = V_{b_2} - \dfrac{Q}{\pi D_2 B_2 \tan \beta_2}$

where $\quad V_{b_2} = \dfrac{\pi D_2 N}{60}.$

From equation (1)

$$H_{\text{mon}} = \frac{V_{w_2} V_{b_2}}{g} = \left(V_{b_2} - \frac{Q}{\pi D_2 B_2 \tan \beta_2} \right) \cdot \left(\frac{V_{b_2}}{g} \right)$$

$$= \frac{Vb_2^2}{g} - \frac{Q V_{b_2}}{\pi D_2 B_2 \tan \beta_2 g}$$

$$= \frac{(\pi D_2 N)^2}{(60)^2 g} - \frac{\pi D_2 N}{60 g} \cdot \frac{Q}{\pi D_2 B_2 \tan \beta_2}$$

Assume $\quad \dfrac{(\pi D_2 N)^2}{(60)^2 g} = C$

and $\quad \dfrac{N}{g B_2 \tan \beta_2} = E$

$$H_{\text{mon}} = C - EQ \qquad ...(2)$$

For a given size of pump and N = constant, N, D_2, B_2 and β_2 are constant.

$\therefore$ E and C are constant.

$\therefore$ Equation (2) is a linear relationship.

(i) For $\beta_2 = \frac{\pi}{2}$, $\cot \beta_2 = 0$.

$\therefore H_{mon} = C$ straight radial blade.

(ii) When $\beta_2 < \frac{\pi}{2}$

H_{mon} decreases with Q. Backward curved vanes

(iii) When $\beta_2 > \frac{\pi}{2}$

H_{mon} increases with Q. Forward curved vanes

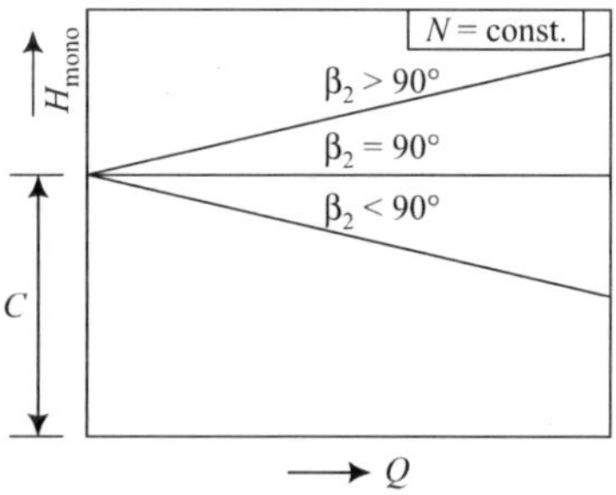

Fig. 9.11 Types of vanes

9.5.3 Performance of Different Vane Shapes

The exit relative velocity V_{r_2} is fixed by β_2, hence the exit velocity triangle depends upon β_2.

1. Radial vanes

Liquid leaves the vane in radial direction as $\beta_2 = 90°$. $V_{w_2} = V_{b_2}$

$$H_{mon} = C - EQ$$

$$H_{mon} = C \quad \text{as} \quad \cot \frac{\pi}{2} = 0$$

$\therefore$ H_{mon} is independent of Q (Fig. 9.11.)

The power input to impeller,

$$P_i = \rho Qg(C - EQ)$$
$$= \rho g CQ - EQ^2$$

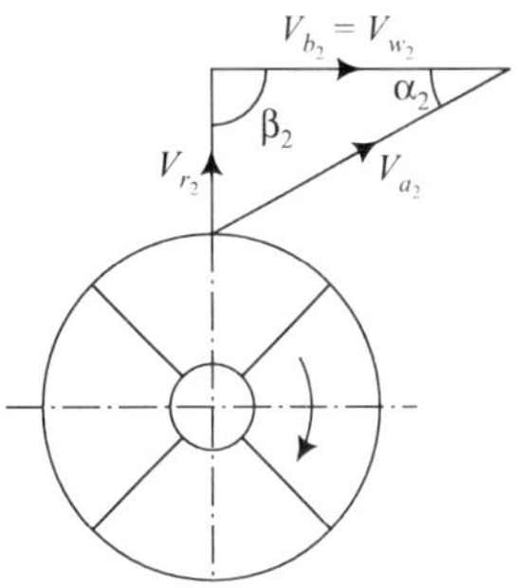

Fig. 9.12 (a) Radial vanes

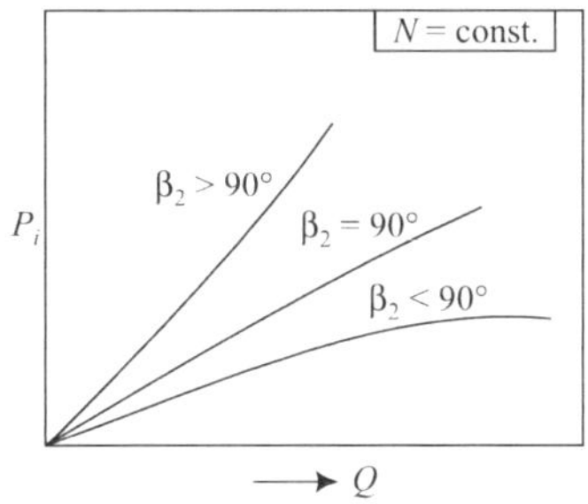

Fig. 9.12 (b) $P_i \propto Q$

For $\beta_2 = 90°$

$$P_i = \rho g Q C$$

$$\therefore \quad P_i \propto Q$$

2. Backward curved vanes

The vane is curved in the opposite direction of rotation and $\beta_2 < 90°$.

$$\cot \beta_2 < 1$$

$$P_i \propto Q H_{\text{mano}}$$

The curve droops downwards.

$$V_{w_2} < V_{b_2}$$

3. Forward curved vanes

The vanes are curved in the direction of rotation of impeller

$$\beta_2 > 90°;\ V_{w_2} > V_{b_2}.$$

$$\cos \beta_2 > 1$$

$$P_i \propto Q\, H_{\text{mano}}$$

The curve rises upwards.

The efficiency increases with decrease of β_2. The relation between maximum impeller efficiency and exit angle β_2 is plotted in Fig. 9.12(e).

For radial blades, $\eta \approx 80\%$

For forward curved blades, $\eta \approx 75\%$

For backward curved blades, $\eta = 85 - 90\%$

The energy conversion efficiency is maximum for forward-curved vanes, but the exit velocity (V_{a_2}) is very high which is not desirable. Normally backward-curved vanes with β_2 between 20° and 30° are used for low and medium head pumps. For high head pumps, forward-curved or radial vane impeller are preferred.

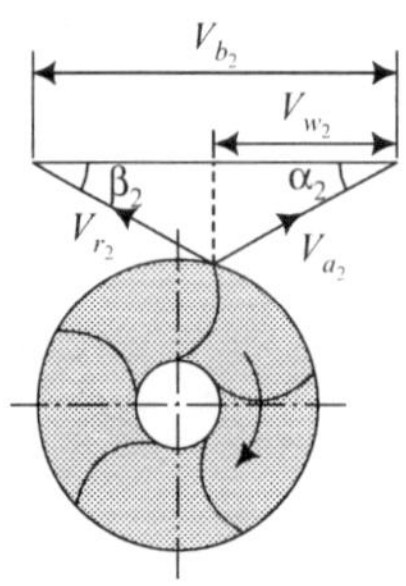

Fig. 9.12 (c) Backward curved vanes

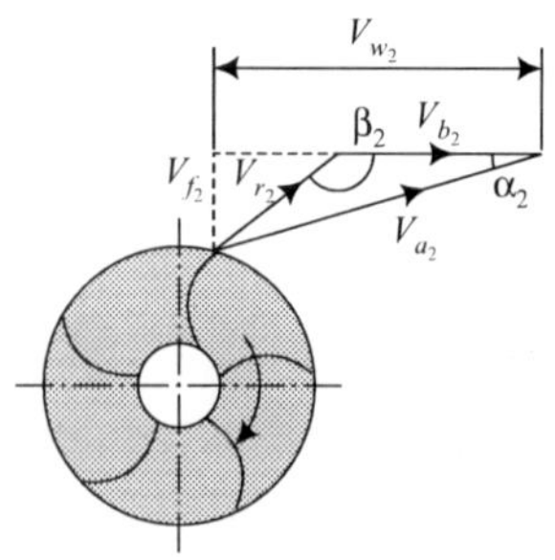

Fig. 9.12 (d) Forward curved vanes

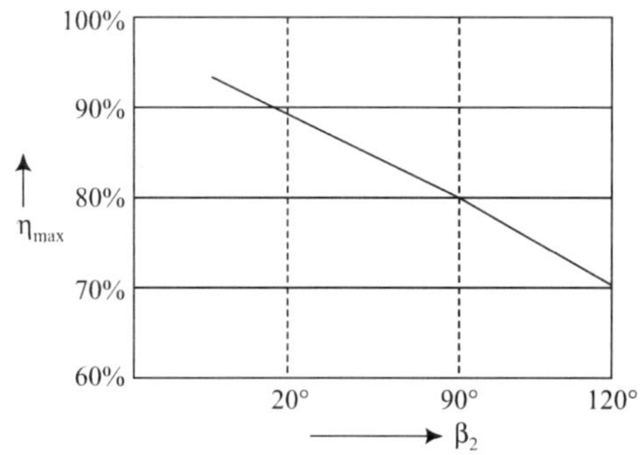

Fig. 9.12 (e) η *v/s* β_2

9.6 DESIGN OF IMPELLER

Design of pump impeller involves the following steps.

9.6.1 Selection of Speed

The rotating speed of impeller is selected to meet the head and capacity requirements. The specific speed and type of impeller is selected.

The selection of speed is governed by the following factors:

1. Type of driving unit
2. *Specific speed.* Higher specific speed results in a small pump and cheaper driver.
3. The optimum hydraulic efficiency and overall efficiency for each type of impeller depends upon the specific speed.
4. In case total required head cannot be produced in one stage, the head is divided between two or more stages. The head per stage also affects its final specific speed and the overall pump efficiency.

5. An impeller model from existing models of same specific speed and hydraulic performance is selected.

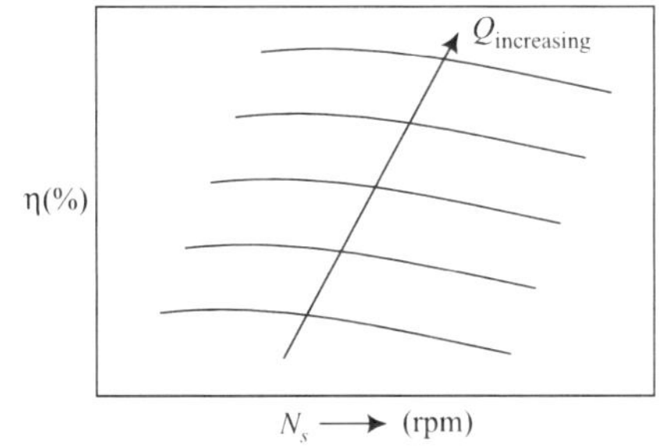

Fig. 9.13 η *v/s* N_S

9.6.2 Impeller Profile

The impeller profile for maximum hydraulic efficiency is determined from the following known data.

1. Inlet and exit velocities of the impeller (V_{a_1} and V_{a_2})
2. Impeller outside diameter and vane velocity (D_2 and V_{b_2})
3. Impeller vane outlet angles (α_2 and β_2) from velocity triangle

In straight radial vanes all liquid streams enter and leave the impeller at the same diameter and the vane is plane of simple profile. The impeller design is determined from one inlet and one outlet velocity triangle.

For mixed flow and axial flow impellers, velocity triangle are drawn for several liquid stream line. Three streamlines are usually sufficient for average mixed flow and axial flow impellers. Variation of vane angles along the radius determines the vane curvature and twist.

9.6.3 Vane Discharge Angle (β_2)

Vane discharge angle is the most important design factor. Theoretical characteristics of pump impeller can be determined from vane angle alone. In actual pumps also, angle β_2 is the deciding factor in impeller design. All design constants are functions of β_2. Proper selection of β_2 is the important step in fixing the design constants of the impeller.

The selection of β_2 depends upon the following factors:

1. The steepness of the head-capacity curve.
2. Normal head and capacity increase with β_2.
3. Maximum capacity is required for a given impeller diameter.

If the above design constraints are not applicable, β_2 is selected from the following considerations:

1. Optimum efficiency and normal design.
2. An average value of $\beta_2 = 22.5°$ is taken for normal impeller design for all specific speeds.
3. For increased output, β_2 should be raised to 27.5° without affecting the efficiency.
4. $\beta_2 = 17.5°$ can be used as lower value for good design.

9.6.4 Mean Effective Impeller Diameter

The various dimensions of an impeller are shown in Fig. 9.14.

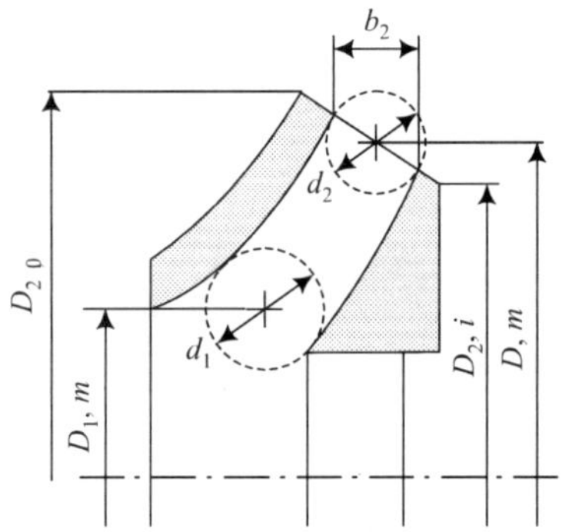

Fig. 9.14 Impeller diameter

For a radial impeller, the mean effective diameter,

$$D_m^2 = \frac{D_2^2, 0 + D_2^2, i}{2}$$

where D_2, 0 and D_2, i are the maximum and minimum outside diameters of impeller.

For an axial flow impeller,

$$D_m^2 = \frac{D_0^2(1 + \nu^2)}{2}$$

where $\nu = \dfrac{D_h}{D_0}$

D_0 = outside diameter of impeller

D_h = hub diameter of impeller

For mixed flow and axial flow impellers, the mean effective diameter divides the liquid flow in the impeller into equal parts.

If the peripheral velocity (V_{b_2}) and impeller discharge angle (β_2) are based on mean effective diameter (D_m), the correlations and plots of several design constants for a complete range of impellers from straight radial to axial flow become very simple.

9.6.5 The Inlet Flow Velocity and Flow Areas

The flow velocity at inlet (V_{f_1}) must be known to complete the impeller profile. This is given by flow ratio (ψ_1)

$$\psi_1 = \frac{V_{f_1}}{\sqrt{2gH}}$$

The leakage at the entrance has been neglected.

The flow velocity at exit (V_{f_2}) of impeller,

$$V_{f_2} = \frac{Q}{A_2} = \frac{Q}{(\pi D_m - \text{Area occupied by vanes})d_2} = \frac{Q}{\pi D_m d_2}$$

$$V_{f_1} = \frac{Q}{A_1} = \frac{Q}{\pi D_{1,m} d_1}$$

The area occupied by vanes is normally neglected.

The dimensions d_1 and d_2 are shown in Fig. 9.14.

For multi-stage pumps such as boiler feed pumps, the flow velocity ratio $\left(\frac{V_{f_1}}{V_{f_2}}\right)$ may be as high as 1.625.

For services where low NPSH requirements are important,

$$V_{f_1} \leq V_{f_2}$$

A low V_{f_1} leads to low entrance angle (α_1).

9.6.6 Reduction of Impeller Diameter

The impeller diameter is generally reduced to decrease the head and capacity of a given centrifugal pump. The performance of pump with reduced impeller diameter can be more or less estimated from similarity laws. However, there are different rules for estimation of performance of different types of pump impellers.

1. Radial impeller

The inlet and outlet velocity triangles for the full and reduced impeller are shown in Fig. 9.15.

$$P_i = \rho Qg\, H_{\text{mono}}$$

$$H_{\text{mono}} = \frac{V_{b_2} V_{w_2} - V_{b_1} V_{w_1}}{g}$$

If the term $\dfrac{V_{b_1} V_{w_1}}{g}$ is ignored, the impeller performance will result in:

1. The head varies directly as the square of diameter ratio and the discharge varies directly as the diameter ratio.
2. The power input varies directly as the cube of the impeller diameter ratio.

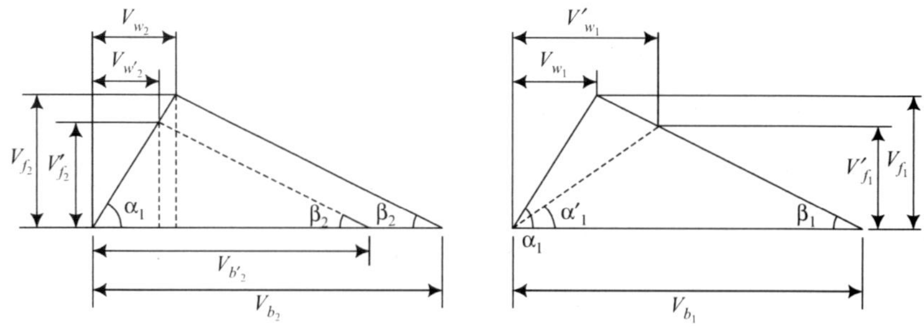

(a) Exit velocity triangles (b) Inlet velocity triangles

Fig. 9.15 Velocity triangles for full and reduced impellers

2. Axial flow impeller

If the impeller diameter is reduced, a new casing or a liner would be required to accommodate the new impeller. Therefore, this practice is rarely in practised.

The impeller performance will result in:

1. The discharge varies directly as the net area swept by the impeller vane, the axial velocity remaining the same.
2. The discharge head will be reduced directly as the square of the diameter ratio.

The new head H' will be:

$$2H' = H_h + H_o\left(\frac{D'_o}{D_o}\right)^2$$

where D'_o = Reduced impeller diameter

D_o = Original impeller diameter

H_o = Original head

H_h = Head at the hub diameter

3. Mixed flow impeller

Mixed flow impeller is difficult to reduce as the vanes overlap which causes great head loss. The diameter ratio is calculated as the average of the diameters of the outside and inside shrouds. Both the diameters should be cut in the same ratio. It is better to reduce the outside diameter and nothing at the hub.

9.6.7 Design Constants

The following constants are used in the design of centrifugal pumps.

1. *Speed ratio* (ϕ) is the ratio of peripheral speed V_{b_2} at the impeller tip and liquid velocity due to theoretical head corresponding to manometric head.

   $$\phi = \frac{V_{b_2}}{\sqrt{2gH_{\text{mano}}}}$$

   ϕ varies between 0.95 and 1.25.

2. *Flow ratio* (ψ) is the ratio of the flow velocity V_{f_2} and liquid velocity due to theoretical manometric head,

   $$\phi = \frac{V_{f_2}}{\sqrt{2gH_{\text{mano}}}}$$

   ψ varies between 0.1 and 0.25.

3. *Diameter ratio* $\left(\frac{D_1}{D_2}\right)$ is the ratio of diameter of impeller at the inner and outer peripheries,

   $$D_2 = \frac{60V_{b_2}}{\pi N}$$

   $\frac{D_1}{D_2}$ varies from $\frac{1}{3}$ to $\frac{2}{3}$.

4. *Exit vane angle* (β_2). It depends upon the head capacity curve. For optimum efficiency $\beta_2 = 25°$ for all values of specific speed.

5. *Inlet vane angle* (α_1) is taken as 90° to ensure radial inlet absolute velocity.

6. *Radius of curvature of vane* is selected to ensure smooth free flow of liquid in the impeller passage. It depends upon inlet and outlet vane angles.

7. *Number of vanes of impeller* depends upon pump size, speed ratio and outlet vane angle.

   With low values of β_2, six or eight vanes are used.

9.7 DYNAMIC SIMILARITY AND PERFORMANCE CHARACTERISTICS

For comparison of characteristics of similar pumps, the following assumptions are made:

1. The pumps are geometrically similar and all linear dimensions are proportional to impeller diameter.
   Example: Impeller width $B \propto$ Impeller diameter D
2. The velocity triangles are similar.
3. The pumps have same hydraulic and volumetric efficiency.
4. All velocities are proportional to $H^{1/2}$.
5. The Reynolds numbers are same for the pumps being compared/examined.

9.7.1 Flow Coefficient

The discharge,

$$Q = \pi D B V_f$$

$$V_f = \psi\sqrt{2gH}$$

$$\propto \sqrt{H}$$

$$B \propto D$$

$$\therefore \quad Q \propto D^2\sqrt{H}$$

Again $$V_b = \frac{\pi D N}{60}$$

$$D = \frac{60V_b}{\pi N}$$

$$\therefore \quad D \propto \frac{V_b}{N}$$

$$\propto \frac{\sqrt{H}}{N}$$

$$V_b \propto ND$$

$$\therefore \quad Q \infty ND^3$$

$$\therefore \quad \frac{Q}{ND^3} = \text{constant} \qquad ...(1)$$

The factor $\frac{Q}{ND^3}$ is called flow coefficient.

9.7.2 Head Coefficient

$$D \propto \frac{V_b}{N}$$

$$\therefore \quad D \propto \frac{\sqrt{H}}{N}$$

$$\sqrt{H} \propto DN$$

$$H \propto D^2N^2$$

$$\therefore \quad \frac{H}{D^2N^2} = \text{constant} \qquad \text{...(2)}$$

The factor $\frac{H}{D^2N^2}$ is called head coefficient or lift coefficient.

9.7.3 Power Coefficient

$$P = \rho QgN$$

$$\propto QH,$$

But $\quad H \propto ND^3$

and $\quad N \propto N^2D^2$

$$\therefore \quad P \propto (ND^3) \times (N^2D^2)$$

$$\infty N^3D^5$$

$$\therefore \quad \frac{P}{N^3D^5} = \text{constant} \qquad \text{...(3)}$$

$\frac{P}{N^3D^5}$ is called power coefficient.

The flow coefficient, head coefficient and power coefficient are used to determine the performance of prototype pumps from the test results of model of pumps. The simplified non-dimensional coefficients can be used for comparison.

$$\frac{Q}{N} = \text{const.}; \quad \frac{H}{N^2} = \text{const.}; \quad \frac{P}{N^3} = \text{const.}$$

9.7.4 Specific Speed

The specific speed of a centrifugal pump is defined as the speed of a geometrically similar pump which would deliver unit quantity (1 m^3/s) against a unit head (1 m).

$$Q \propto D^2\sqrt{H}$$

$$D^2 \propto \frac{H^2}{N^2}$$

$$\therefore \quad Q \propto \frac{H^2}{N^2}\sqrt{H}$$

$$\propto \frac{H^{3/2}}{N^2}$$

$$\therefore \quad Q = K\frac{H^{3/2}}{N^2}$$

where K = constant of proportionality

when $H = 1$ m and $Q = 1$ m/s^2, $N = N_s$

$$1 = K\frac{(1)^{3/2}}{N_s^2}$$

$$\therefore \quad K = N_s^2$$

$$\therefore \quad Q = N_s^2\frac{H^{3/2}}{N^2}$$

Specific speed,

$$N_s = \frac{N\sqrt{Q}}{H^{3/2}}$$

Notes:

1. For multi-stage pumps, H is the manometric head per stage.
2. For double suction pump, Q is the half of total discharge.

9.7.5 Relation Between Specific Speed of Pump and Turbine

$$(N_s)_{\text{turbine}} = \frac{N\sqrt{P}}{H^{5/4}}$$

$$(N_s)_{\text{pump}} = \frac{N\sqrt{Q}}{H^{3/4}}$$

$$P = \frac{\rho QgH}{1000}\eta_o \quad \text{[kW]}$$

$$\therefore \quad (N_s)_{\text{turbine}} = N\left(\frac{\rho QgH}{1000}\eta_o\right)^{1/2} \times \frac{1}{H^{5/4}}$$

$$(N_s)_{\text{turbine}} = \left(\frac{\rho g \eta_o}{1000}\right)^{1/2} \cdot \frac{N\sqrt{Q}}{H^{3/4}}$$

$$= \left(\frac{\rho g \eta_o}{1000}\right)^{1/2} \times (N_s)_{\text{pump}}$$

Assume η_o for a turbine as 90% and ρ for water as 1000 kg/m^2

$$(N_s)_{\text{turbine}} = \left(\frac{1000 \times 9.81 \times 0.9}{1000}\right)^{1/2} \times (N_s)_{\text{pump}}$$

$$= 2.97(N_s)_{\text{pump}}$$

9.7.6 Performance Characteristics

Performance characteristics of a pump are variation of head, power and efficiency against flow rate graphically. There are three types of performance characteristics.

1. Main characteristics

The pump is run at a constant speed and discharge is varied. Measurements are made for each discharge. The values of head and shaft power are measured by pressure gauge and dynamometer respectively and calculations are made for pump efficiency.

A set of curves are plotted for different constant speeds.

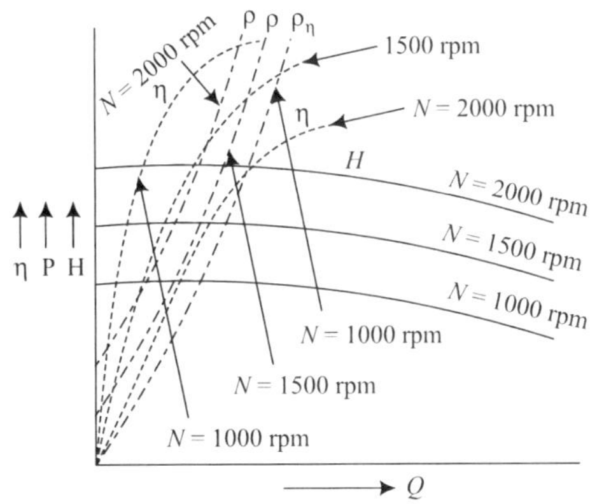

Fig. 9.16 Main characteristics

2. Operating characteristics

The operating characteristics are the main characteristics for design speed for which maximum efficiency is obtained. Operating characteristics are used to obtain design discharge and head for maximum efficiency.

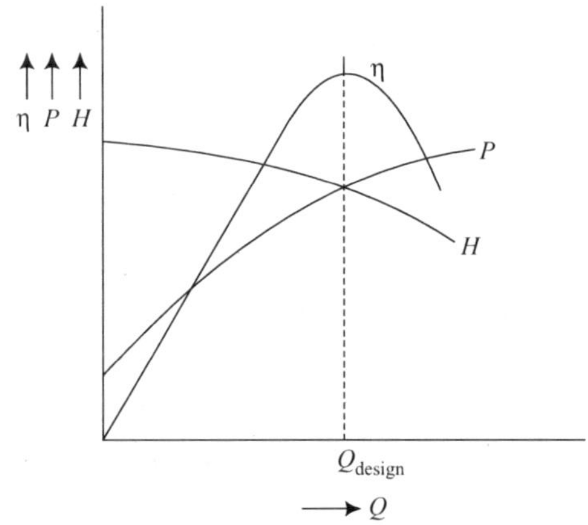

Fig. 9.17 Operating characteristics

3. Universal characteristics

These are constant efficiency curves also called Muschel curves. These curves are plotted from the main characteristics for different constant efficiency values. These curves help the region where pump would operate with maximum efficiency.

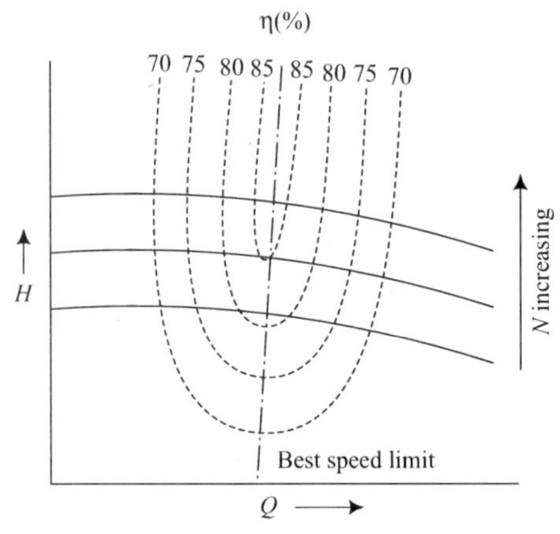

Fig. 9.18 Universal characteristics

9.8 CAVITATION AND MAXIMUM SUCTION LIFT

Cavitation starts under the vanes at pump entry where the pressure is the lowest. If the pressure is lower than vapour pressure of the liquid at the prevailing temperature, there is separation of water and bubbles are formed. These bubbles are carried to high pressure region near the impeller exit. The bubbles collapse causing pitting and severe damage of the metal surface of vane.

The vane tips at exit experience the following harmful effects:

1. Pitting and erosion of surface due to continuous hammering action of collaping bubbles.
2. Sudden drop in head, efficiency and power delivered to the liquid.
3. Noise and vibration produced by the collapse of vapour bubbles.

The following factors contribute to cavitation.

1. High impeller speed
2. Restricted suction area
3. High specific speed
4. High liquid temperature
5. Required NPSH equal to or greater than available NPSH.

The pressure balance at free liquid surface (1) and pump suction (2) gives:

$$H_1 = H_2 + H_{f_s} + H_s + \frac{V_s^2}{2g}$$

where, $H_1 = H_{atm}$ and depends upon site altitude
H_{fs} = Head loss in suction pipe
H_s = Static head between points 2 and 1.
V_s = Liquid velocity through suction pipe
H_2 = Head at suction eye.

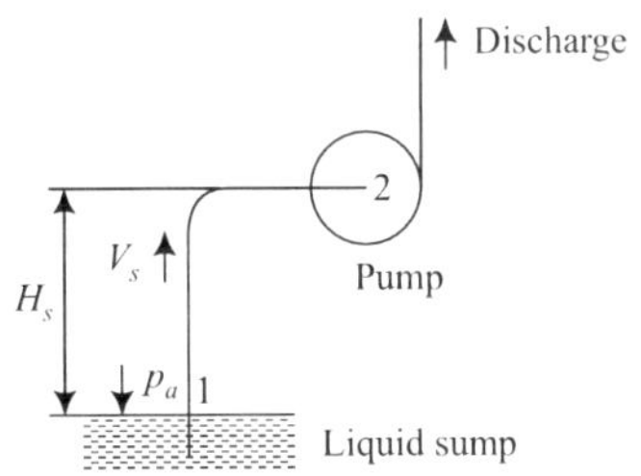

Fig. 9.19 Pressure balance at pump suction

H_2 should be more than vapour pressure of liquid which is a function of temperature.

$$H_2 \geq H_{vap.}$$

If $H_2 < H_{vap}$, water separation occurs resulting in cavitation and harmful effects listed above.

Minimum Net Position Suction Head (NPSH) should be provided to avoid cavitation.

Net Positive Suction Head (NPSH)

$$= H_{\text{atm}} - \left(H_s + H_{f_s} + H_{\text{vap}} + \frac{V_s^2}{2g} \right)$$

$$\geq 2 \text{ to } 2.5 \text{ m}$$

$$\frac{V_s^2}{2g} = H_{\text{atm}} - H_s - H_{fs} - H_{\text{vap}}$$

$$\therefore \quad \frac{V_s^2}{2gH} = \frac{H_{\text{atm}} - H_s - H_{f_s} - H_{\text{vap}}}{H}$$

where H = Head developed by the pump.

The factor $\frac{V_s^2}{2gH}$ is called the Thomas cavitation number (σ) and $[(H_{\text{atm}} - H_s - H_{f_s}) - H_{\text{vap}}]$ represents NPSH, i.e. the head to make the liquid flow through the suction pipe to the pump impeller.

$$\therefore \quad \sigma = \frac{[(H_{atm} - H_s - H_{f_s}) - H_{\text{vap}}]}{H} = \frac{\text{NPSH}}{H}$$

The cavitation factor is a function of specific speed. The critical cavitation number,

$$\sigma_{cr} = 1.042 \times 10^{-3} (N_s)^{4/3}$$

$$\sigma \geq \sigma_{cr}$$

or $$(\text{NPSH}) \geq \sigma_{cr} H.$$

For geometrically similar machines

$$Q \propto ND^3$$

$$\text{NPSH} \propto \frac{V^2}{2g}$$

$$\propto \frac{Q^2}{D^4}$$

or $$\frac{N\sqrt{Q}}{(\text{NPSH})^{3/4}} = \text{const.} = S_{su}$$

where S_{su} = suction specific speed

$$\frac{\text{NPSH}}{H} = \left(\frac{N_s}{s_{su}}\right)^{4/3}$$

or
$$\sigma = \left(\frac{N_s}{s_{su}}\right)^{4/3} \quad \text{For } \sigma = \sigma_{cr}, S_{su} = 175$$

For a given head and capacity, a pump with low specific speed operates safely with greater suction lift.

9.9 MULTISTAGE PUMPS

A number of impellers are connected in series to build up high pressure. Two or more identical impellers are mounted on the same shaft and enclosed in the same casing. The discharge from first impeller passes through a guided passage and enters the second impeller, and discharge from second impeller enters the third impeller, and so on. The discharge from last impeller is pushed through the delivery pipe.

The total discharge is same through each impeller. The total head developed by the pump,

$$H_{\text{total}} = nH_{\text{stage}}$$

where n = Number of impellers or stages

H_{stage} = Head developed by each impeller

$$N_S = \frac{N\sqrt{Q}}{(H/n)^{3/4}}$$

For centrifugal pump, $N_s = 40$ will give an efficient and economical pump.

$$H_{\text{mano}} = \frac{V_{b_2} V_{w_2}}{g}$$

Peripheral speed (V_{b_2}) should be increased to get more head. However, V_{b_2} is limited by impeller strength and cavitation. V_{b_2} is limited to 40 m/s for C.I. impeller and 300 m/s for alloy steel impellers.

9.10 PUMP OPERATION

9.10.1 Operational Problems

The main operational problems of centrifugal pumps along with their causes are tabulated in Table 9.2.

Table 9.2 Operational problems of centrifugal pumps

S.No.	Problem	Reasons
1.	Pump fails to deliver liquid.	(i) Pump not properly primed (ii) Suction lift and delivery too high (iii) Speed low (iv) Wrong direction of pump rotation (v) Clogging of impeller, strainer or suction line
2.	Not full head and capacity	(i) Air leakage into pump (ii) Speed low (iii) Wrong direction of pump rotation (iv) Partial clogging of impeller (v) Small impeller diameter (vi) Discharge head higher than anticipated (vii) Insufficient suction head when pumping hot or volatile liquids
3.	Power consumption by pump excessive	(i) Pump running on higher speed (ii) Wrong direction of pump rotation (iii) Low head and high discharge (iv) Bent shaft and rubbing of runner with casing (v) Liquid density high (vi) Cavitation (vii) Lack of lubrication (viii) Tight packing
4.	Overheating of pump	(i) Rubbing of runner with casing (ii) Lack of lubrication (iii) High impeller speed (iv) Worn-out rings, bent shaft, uncleaned bearings or other mechanical defects
5.	Noise and vibration	(i) Cavitation (ii) Misalignment of pump shaft (iii) Worn-out bearings (iv) Improper foundation (v) Rubbing of runner with casing

9.10.2 Priming of Pump

When the centrifugal pump is not primed (filled with liquid) before starting, the air pockets inside the impeller may give rise to formation of vortices and cause discontinuity of liquid flow. Priming is done by pouring of liquid into pump casing (when pump is not running) through a funnel and air is displaced through air vents provided on the casing.

The problem of formation of air pockets can be avoided by the following methods to ensure proper delivery of liquid by the pump when running.

1. Installation of pump below suction liquid level
2. Self priming device
3. Priming of pump

1. Self priming devices

(i) Vacuum pump is run on the same pump shaft which sucks air from suction pipe and throws into the atmosphere.

(ii) Proprietary design of pump to form air-liquid mixture.

2. Priming methods

(i) Manually pouring liquid through priming funnel and opening vent valves till pump is fully filled. The liquid is poured when pump is not running.

(ii) Connecting the pump with city water mains for filling water into impeller and suction pipe.

(iii) Priming chamber is provided on delivery side which is filled with liquid when pump is running. It is used for priming. The size of priming chamber is three times the volume of suction pipe.

9.10.3 Minimum Starting Speed

The centrifugal pump starts delivering liquid when the following conditions are fulfilled.

(i) There is no air entrapped on the suction side of pump. This is ensured by priming the pump which consists of filling liquid into impeller, casing and suction pipe in order to remove air, gas or vapour before starting.

(ii) The pressure rize in the impeller is higher than the gross or manometric head of the pump.

The centrifugal head or pressure head created the pump is $\dfrac{V_{b2}^2 - V_{b1}^2}{2g}$.

The manometric head,

$$H_{\text{mano}} = \frac{V_{b_2} V_{w_2}}{g} . \eta_{\text{mano}}$$

$$\therefore \qquad \frac{V_b^2 - V_{b1}^2}{2g} \geq \frac{V_{b_2} V_{w_2}}{g} . \eta_{\text{mano}} \text{ to ensure delivery of liquid}$$

For η_{mano} to be 100%

$$\frac{V_{b_2} V_{w_2}}{g} = \frac{\omega^2}{2g} (R_2^2 - R_1^2)$$

$$\frac{V_{b_2} V_{w_2}}{g} = \left(\frac{2\pi N}{60}\right) \cdot \frac{(R_2^2 - R_1^2)}{2g}$$

where R_1 and R_2 are impellers inside and outside the radii.

The impeller speed N (rpm) can be found out. In other words, R_2 can be found out which will ensure delivery of liquid at its nominal rotational speed.

9.10.4 Effect of Speed and Diameter

The head and capacity of an existing pump can be adjusted to suit site conditions by altering the pump speed or impeller diameter or both.

1. Impeller speed

When the diameter of impeller is not changed, the performance of pump can be altered by changing the pump speed.

$$V_b = \frac{\pi D N}{60}$$

$$\therefore \qquad V_b \propto N$$

For same vane angles β_1, β_2, α_1 and α_2, the flow velocity

$$V_f \propto V_b \propto N$$

and

$$Q \propto V_f \propto N.$$

$$\therefore \qquad \frac{Q}{Q'} = \frac{N}{N'} \qquad \text{...(1)}$$

The flow of liquid is directly proportional to pump speed.

$$V_b \propto \sqrt{H}$$

$$\therefore \qquad H \propto V_b^2 \propto N^2$$

$$\therefore \qquad \frac{H}{H'} = \left(\frac{N}{N'}\right)^2 \qquad \text{...(2)}$$

The discharge head is proportional to the square of speed.

$$P \propto QH$$

$$\therefore \qquad \frac{P}{P'} = \left(\frac{N}{N'}\right)^3 \qquad \text{...(3)}$$

The power input to pump is proportional to the cube of the speed.

2. Impeller diameter

When the speed of the impeller is not changed, the pump performance can be adjusted by changing the impeller diameter either by cutting or using an impeller of different diameter in the same pump casing.

The vane speed,

$$V_b = \frac{\pi DN}{60}$$

$$\therefore \qquad V_b \propto D \qquad (\because N = \text{constant})$$

The flow velocity,

$$V_f \propto V_b \propto D$$

The flow area,

$$A_f \propto D^2$$

The liquid flow,

$$Q \propto A_f V_f$$

$$\propto D^3$$

$$\frac{Q}{Q'} = \left(\frac{D}{D'}\right)^3 \qquad \text{...(4)}$$

The pump capacity is proportional to the cube of the impeller diameter.

The pump head,

$$H \propto V_b^2$$

$$\propto = D^2$$

$$\frac{H}{H'} = \left(\frac{D}{D'}\right)^2 \qquad \text{...(5)}$$

The pump head is proportional to the square of the impeller diameter.

The pump power,

$$P \propto QH$$

$$\therefore \qquad \frac{P}{P'} = \left(\frac{D}{D'}\right)^5 \qquad \text{...(6)}$$

The power consumption changes by 5th power of impeller diameter.

3. Impeller diameter and speed

The impeller diameter and speed are changed simultaneously

From equations (1) and (4),

$$\frac{Q}{Q'} = \left(\frac{N}{N'}\right)\left(\frac{D}{D'}\right)^3$$

From equations (2) and (5)

$$\frac{H}{H'} = \left(\frac{N}{N'}\right)^2\left(\frac{D}{D'}\right)^2$$

From equations (3) and (6)

$$\frac{P}{P'} = \left(\frac{N}{N'}\right)^3\left(\frac{D}{D'}\right)^5$$

QUESTION BANK NO. 9

1. Discuss the classification of centrifugal pumps as per specific speed, working head, pump construction and the liquid handled.
2. Draw the installation of a centrifugal pump and discuss the significance of various heads.
3. Distinguish between manometric efficiency and hydraulic efficiency of a centrifugal pump.
4. With the help of an energy flow diagram, discuss various losses and efficiencies of a centrifugal pump.
5. Draw the velocity diagrams of centrifugal pump. What is the effect of vane shape on velocity triangles?
6. Discuss the influence of exit blade angle on the performance and efficiency of a centrifugal pump.
7. Explain step-by-step the procedure for the design of a centrifugal pump.
8. Define important ratios and coefficients as applicable to the hydraulic design of a centrifugal pump.
9. Explain various performance characteristics of a pump under the headings:
   (a) Main characteristics
   (b) Operating characteristics
   (c) Universal characteristics
10. What is cavitation? What is NPSH? How are cavitation and NPSH correlated?

11. Write notes on:
    (a) Priming of pumps
    (b) Minimum starting speed
    (c) Effect of speed and diameter on pump performance.
12. Prepare a list of problems and reasons in the malfunctioning of a centrifugal pump.

TUTORIAL SHEET NO. 9

1. Calculate the head imparted to a liquid and inlet vane angle it passes through an impeller of 0.25 m diameter and 0.1 m inlet diameter rotated at 1440 rpm. The outlet vane angle is set back at an angle of 20° to the tangent. Assume radial entrance and velocity of flow at 3 m/sec.

Solution The data given:

$$D_1 = 0.10 \text{ m}$$

$$D_2 = 0.25 \text{ m}$$

$$N = 1440 \text{ rpm}$$

$$\alpha_1 = 90°$$

$$\beta_2 = 20°$$

$$V_{f_1} = V_{f_2} = 3 \text{ m/s}$$

The vanes are backward curved with velocity triangles as shown.

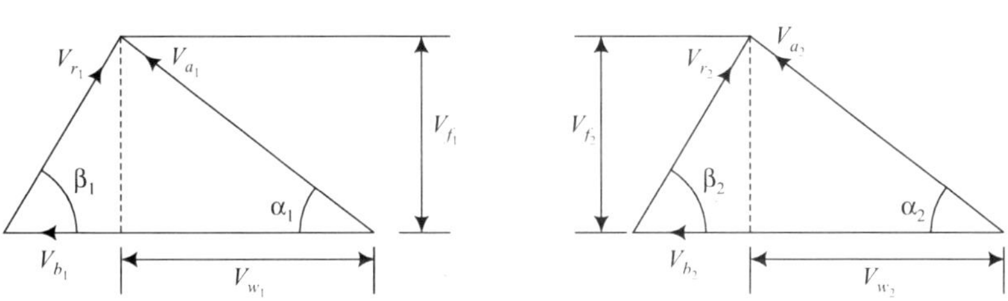

The vane velocity at inlet,

$$V_{b_1} = \frac{\pi D_1 N}{60} = \frac{\pi \times 0.1 \times 1440}{60} = 7.54 \text{ m/s}$$

$$V_{b_2} = \frac{\pi D_2 N}{60} = \frac{\pi \times 0.25 \times 1440}{60} = 18.85 \text{ m/s}$$

From exit velocity triangle,

$$V_{b_2} - V_{w_2} = \frac{V_{f_2}}{\tan \beta_2} = \frac{3}{\tan 20°} = 8.24 \text{ m/s}$$

$\therefore \qquad V_{w_2} = 18.85 - 8.24 = 10.6$ m/s

(i) Head imparted to liquid

$$H = \frac{V_{b_2} V_{w_2}}{g} = \frac{18.85 \times 10.6}{9.81} = \mathbf{20.37\ m} \quad \text{Ans}$$

(ii) From inlet velocity triangle,

$$\tan \beta_1 = \frac{V_{f_1}}{V_{b_1}} \qquad (\because \alpha_1 = 90°)$$

$$= \frac{3}{7.54}$$

$$\therefore \qquad \beta_1 = \mathbf{21.7°} \quad \text{Ans}$$

2. A centrifugal pump impeller has an external diameter of 0.4 m and internal diameter of 0.18 m and it runs at 1440 rpm. Assuming a constant radial flow through the impeller at 2.5 m/sec and the vanes at exit are set back at an angle of 25°, calculate:
   (a) The angle absolute velocity of water at exit makes with the tangent,
   (b) Inlet vane angle, and
   (c) Work done per kg of water.

Solution The data given:

$$D_1 = 0.18 \text{ m}$$

$$D_2 = 0.40 \text{ m}$$

$$N = 1440 \text{ rpm}$$

$$V_{f_1} = V_{f_2} = 2.5 \text{ m/s}$$

$$\beta_2 = 25°$$

(a)

$$V_{b_1} = \frac{\pi D_1 N}{60} = \frac{\pi \times 0.18 \times 1440}{60} = 13.57 \text{ m/s}$$

$$V_{b_2} = \frac{\pi D_2 N}{60} = \frac{\pi \times 0.4 \times 1440}{60} = 30.16 \text{ m/s}$$

$$\tan \beta_1 = \frac{V_{f_1}}{V_{b_1}} = \frac{2.5}{13.57}$$

$$\therefore \qquad \beta_1 = \mathbf{10.4°}.$$

(b) $$\tan \beta_2 = \frac{V_{f_2}}{V_{b_2} - V_{w_2}} = 0.4663$$

$$\therefore \quad V_{w_2} = \frac{-2.5}{0.4663} + 30.16 = 24.8 \text{ m/s}$$

$$\tan \alpha_2 = \frac{V_{f_2}}{V_{w_2}} = \frac{2.5}{24.8}$$

$\therefore \quad \alpha_2 = \mathbf{5.75°}$ Ans

(c) Work done per kg $= V_{w_2} V_{b_2} = 24.8 \times 30.16$

$= \mathbf{748\ N.m/kg}$ Ans

3. A centrifugal pump in a dock pumps 1565 litres/sec against a mean lift of 6.1 m. The impeller has a diameter of 1.22 m; speed 200 rpm; area at the outer periphery 0.645 m^2; angle of vane at outlet 26° and ratio of external to internal diameter 2:1. Determine:
   (i) the hydraulic efficiency,
   (ii) power required to drive the pump.

   Solution The given data:

$$Q = 1.565 \text{ m}^3\text{/s}$$

$$H = 6.1 \text{ m}$$

$$D_2 = 1.22 \text{ m}$$

$$N = 200 \text{ rpm}$$

$$A_f = 0.645 \text{ m}^2$$

$$\beta_2 = 26°$$

$$D_1 = \frac{D_2}{2} = \frac{1.22}{2} = 0.61 \text{ m.}$$

$$V_{f_2} = V_{f_1} = \frac{Q}{A_f} = \frac{1.565}{0.645} = 2.43 \text{ m/s}$$

(i) $$V_{b_1} = \frac{\pi D_1 N}{60} = \frac{\pi \times 0.61 \times 200}{60} = 6.39 \text{ m/s}$$

$$V_{b_2} = \frac{\pi D_2 N}{60} = \frac{\pi \times 1.22 \times 200}{60} = 12.78 \text{ m/s}$$

$$\eta_{hyd} = \frac{\rho Q g H}{\rho Q g V_{w_2} V_{b_2}} = \frac{Hg}{V_{w_2} V_{b_2}}$$

$$\tan \beta_2 = \frac{V_{f_2}}{V_{b_2} V_{w_2}} = 0.4877$$

$$\therefore \quad V_{w_2} = \frac{2.43}{0.4877} - 12.78 = 7.83 \text{ m/s}$$

$$\eta_{hyd} = \frac{6.1 \times 9.81}{7.83 \times 12.78} = 0.598 \approx \mathbf{60\%} \quad \text{Ans}$$

(ii) The power required to drive the pump,

$$P_i = \frac{\rho Q g H}{\eta_{hyd}} = \frac{1000 \times 9.81 \times 1.65 \times 6.1 \times 10^{-3}}{0.6}$$

$$= \mathbf{156\ kW} \quad \text{Ans}$$

4. A centrifugal pump lifts water under a static head of 33.6 m of which 3.68 m is suction lift. Suction and delivery pipes are both 1.5 m in diameter. The head loss in suction pipe is 2.1 m, and in the delivery pipe is 7.1 m. The impeller is 0.38 m in diameter and 2.54 cm width and revolves at 0.38 m in diameter and 2.54 cm wide and revolves at 1200 rpm. Its exit blade angle is 35°. If the manometric efficiency of the pump is 82%, determine the discharge and the pressure at the suction and delivery branches of the pump.

Solution The given data

$$H_{stat} = 33.6 \text{ m} \qquad \eta_{mano} = 82\%$$

$$H_S = 3.68 \text{ m}$$

$$H_d = 33.6 - 3.68 = 29.92 \text{ m}$$

Total head to be supported by the pump,

$$H_{mano} = H_{static} + \sum \Delta H_L$$

$$= 33.6 + 2.1 + 7.1 = 42.8 \text{ m of } H_2O.$$

$$V_{b_2} = \frac{\pi D_2 N}{60} = \frac{\pi \times 0.38 \times 1200}{60} = 23.88 \text{ m/s}$$

$$\eta_{mano} = \frac{H_{mano}}{V_{b_2} V_{w_2} / g}$$

$$\therefore \qquad V_{w_2} = \frac{H_{\text{mano}} \cdot g}{V_{b_2} \eta_{\text{mano}}} = \frac{42.8 \times 9.81}{23.88 \times 0.82}$$

$$= 21.44 \text{ m/s}$$

$$\tan \beta_2 = \frac{V_{f_2}}{V_{b_2} - V_{w_2}}$$

$$\therefore \qquad V_{f_2} = (V_{b_2} - V_{w_2}) \tan \beta_2 = (23.88 - 21.44) \tan 35°$$

$$= 1.7 \text{ m/s}$$

$$Q = \pi D_2 B_2 V_{f_2}$$

$$= \pi \times 0.38 \times \frac{0.254}{100} \times 1.7 = 0.00515 \text{ m}^3\text{/s}$$

Velocity in suction and delivery pipes,

$$V_d = V_s = \frac{Q}{A_s} = \frac{0.00515}{\pi/4 \times (0.15)^2} = 0.29 \text{ m/s}$$

$$\text{Velocity head} = \frac{V_s^2}{2g} = \frac{(0.29)^2}{2 \times 9.81} = 0.043 \text{ m}$$

Total effective pressure on delivery side

$$H_d + H_{L_d} + \frac{V_d^2}{2g} = 29.92 + 7.1 + 0.043$$

$$= \mathbf{37\ m}$$

Total effective pressure or suction side

$$= H_s + H_{L_s} + \frac{V_s^2}{2g} = 3.68 + 2.1 + 0.043$$

$$= 5.82 \text{ m of vacuum.}$$

or $\qquad (10 - 5.82) = \mathbf{4.177\ m}$ absolute

5. A centrifugal pump discharges 25 litres per second against a total head of 15 m at 1400 rpm and the diameter of the impeller is 0.45 m. A geometrically similar pump of 30 cm diameter is run at 2800 rpm. Calculate the head, discharge and power ratio required, assuming equal efficiencies between the two pumps.

Solution The given data for first pump:

$$Q_1 = 25 \ l/s = 0.025 \text{ m}^2\text{/s}$$

$$H_1 = 15 \text{ m}$$
$$N_1 = 1400 \text{ rpm}$$
$$D_1 = 0.45 \text{ m}$$

The data for geometrically similar pump:

$$D_2 = 0.3 \text{ m}$$
$$N_2 = 2800 \text{ rpm}$$

$$\frac{Q_1}{Q_2} = \left(\frac{N_1}{N_2}\right)\frac{D_1^3}{D_2^3} = \frac{1400}{2800}\left(\frac{0.45}{0.3}\right)^3 = 1.6875$$

$$\frac{H_1}{H_2} = \left(\frac{N_1}{N_2}\right)^2\left(\frac{D_1}{D_2}\right)^2 = \left(\frac{1400}{2800}\right)^2\left(\frac{0.45}{0.3}\right)^2 = 0.5625$$

$$\frac{P_1}{P_2} = \left(\frac{N_1}{N_2}\right)^3\left(\frac{D_1}{D_2}\right)^5 = \left(\frac{1400}{2800}\right)^3\left(\frac{0.45}{0.3}\right)^5$$

$$= 0.125 \times 7.59375 = \mathbf{0.949}$$

6. Determine the number of stages required for a multistage centrifugal pump to deliver 75.7 lit/sec to a height of 91 m at 1500 rpm. Assume a specific speed of 30 rpm and an efficiency of 75%.

Solution The given data

$$Q = 75.7 \text{ lit/sec} = 0.0757 \text{ m}^3\text{/sec}$$
$$H = 91 \text{ m}$$
$$N = 1500 \text{ rpm}$$
$$\eta = 75\%$$
$$N_S = 30$$

$$N_S = \frac{N\sqrt{Q}}{H^{3/4}}$$

$$\therefore \quad H^{3/4} = \frac{N\sqrt{Q}}{N_s} = \frac{1500\sqrt{0.0757}}{30}$$

$$= 10.7568$$

$$\therefore \quad H = 32.93 \text{ m} \quad \text{per stage}$$

No. of stages $n = \dfrac{H}{H_{stage}} = \dfrac{91}{32.93}$

$= 2.76$ say **3**

7. A 1/5 scale model of a centrifugal pump absorbs 20 kW when pumping against a head of 8 m at its best speed of 400 rpm. If the actual pump works against 32 m head, find the speed and power required by the actual pump. Determine also the quantity of water discharged by the two pumps.
[N_a = 160 rpm; P_a = 4000 kW; Q_m = 225 m^3/s; Q_a = 17750 m^3/s].

CHAPTER 10

Reciprocating Pump

10.1 INTRODUCTION

A reciprocating pump is a piston and cylinder arrangement and the liquid follows the movement of piston during suction and delivery strokes. It is a "positive" displacement pump. These are suitable for small capacities and high heads. The capacity is a function of speed and not head. Fig. 10.1 shows the regions of applications as compared to centrifugal and axial flow pumps.

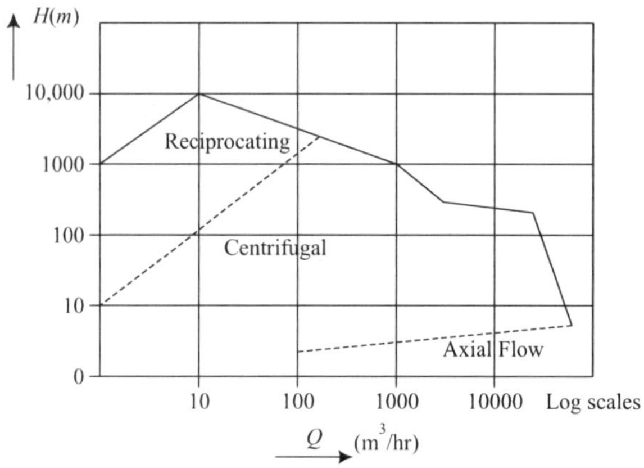

Fig. 10.1 Capacity-head region

It consists of the following:

1. A piston moves backward and forward in a closely fitted cylinder.
2. Suction and delivery pipes with non-return valves.
3. Crank and connecting rod mechanism operated by a prime mover (engine or electric motor). When piston moves outward, low pressure is formed in the cylinder left space. The suction valve opens and the liquid is sucked into the cylinder from the sump. When the piston moves inward, high pressure is developed in the cylinder. The suction valve closes and delivery

valve opens. The liquid is forced into the discharge tank through delivery pipe. The suction and delivery strokes are executed alternately and the liquid is pumped from sump to delivery tank.

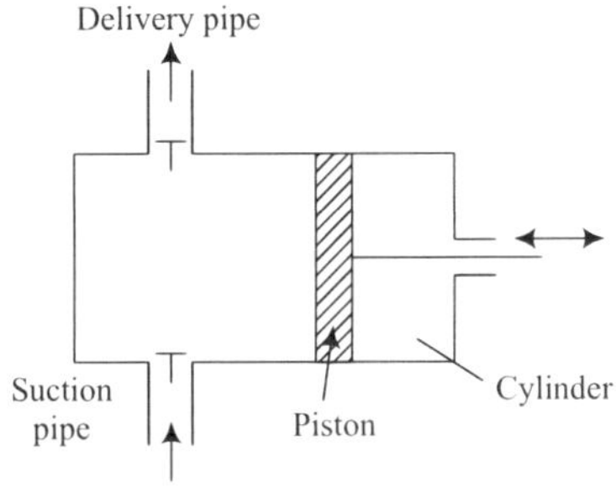

Fig. 10.2 Pump operation

10.2 PUMP INSTALLATION AND PERFORMANCE

The pump layout is shown in Fig. 10.3.

Static head, $H = H_S + H_d$ (m)

Piston area, $A = \frac{\pi}{4} D^2$ (m^2)

Piston rod area, $a = \frac{\pi}{4} d^2$ (m^2)

Piston stoke = L (m)

Crank speed = N rpm

For a single acting pump, the liquid discharge,

$$Q = \frac{ALN}{60} \text{ [m}^3\text{/s]}$$

The equivalent power output,

$$P_o = \rho QgH \text{ [W]}$$

$$= \frac{\rho g}{1000} \cdot \frac{ALN}{60} (H_S + H_d) \text{ [kW]}$$

The overall pump efficiency,

$$\eta_{\text{overall}} = \frac{\rho QgH}{\sqrt{3} EI \cos \phi} = \frac{P_o}{P_E}$$

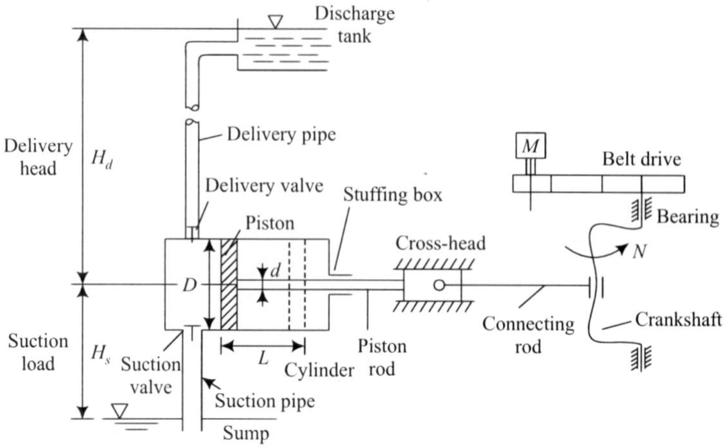

Fig. 10.3 Pump layout

where P_E = Electrical power input to motor.

Force acting on piston during forward stroke

$$F_S = \rho g H_S A \text{ [N]}$$

Force acting on piston during backward stroke

$$F_d = \rho g H_d A \text{ [N]}$$

The rate of delivery of pump,

$$Q = A_S V_S \text{ [m}^3\text{/s]}$$

where A_S = area of suction pipe [m^2]

V_S = velocity of the liquid in suction pipe [m/s]

$$\therefore \quad Q = A_S \omega r \sin\theta \frac{A}{A_S}$$

$$\therefore \quad Q = f(\sin\theta)$$

Note *In double acting pump.*

$$Q = \frac{ALN}{60} + \frac{(A-a)LN}{60} \text{ [m}^3\text{/s]}$$

$$= \frac{LN}{60}(2A - a)$$

$$\approx \frac{2ALN}{60} \qquad (\because a << A \text{ and may be neglected})$$

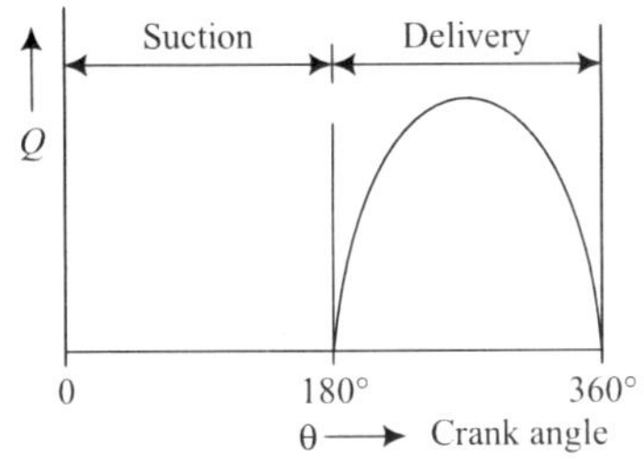

Fig. 10.4 Liquid delivery

$$P_o = \frac{\rho Q g H}{1000} \text{ [kW]}$$

The force on piston during forward stroke,

$$F_S = \rho g H_S A + \rho g H_d (A - a)$$

The force on piston during backward stroke,

$$F_d = \rho g H_S (A - a) + \rho g H_d A$$

10.3 CLASSIFICATION OF RECIPROCATING PUMPS

The flow from a single-cylinder and single acting reciprocating pump, working with simple harmonic motion (Fig. 10.4) is of pulsating and intermittent nature. However, it is desirable to have a continuous flow both during the suction and delivery strokes. This is accomplished by using double acting pump, multi-cylinder pump and pump fitted with air vessels. Accordingly, reciprocating pumps are classified as:

1. Single cylinder pump
   (i) Single acting
   (ii) Double acting
2. Two-cylinder pump or two-throw pump
3. Three-cylinder pump or three-throw pump
4. Pump without air vessel
5. Pump with air vessel

The schematic diagrams and performance curves of various types of pumps are shown in Fig. 10.5.

1. Single cylinder single acting pump.

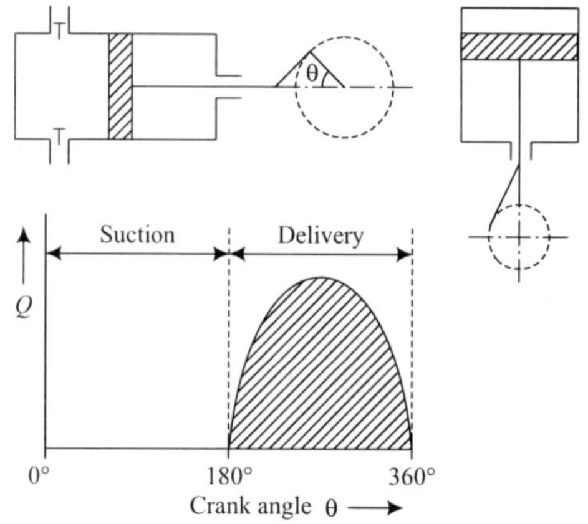

Fig. 10.5 (a) Single cylinder single acting pump

2. Single cylinder double acting pump.

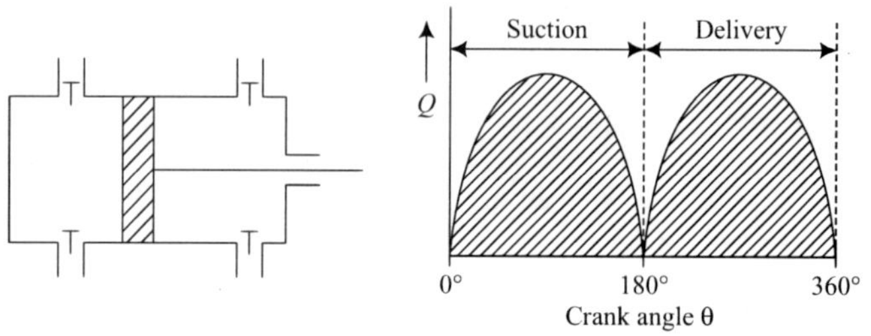

Fig. 10.5 (b) Single cylinder double acting pump

3. Two-cylinder double acting pump.

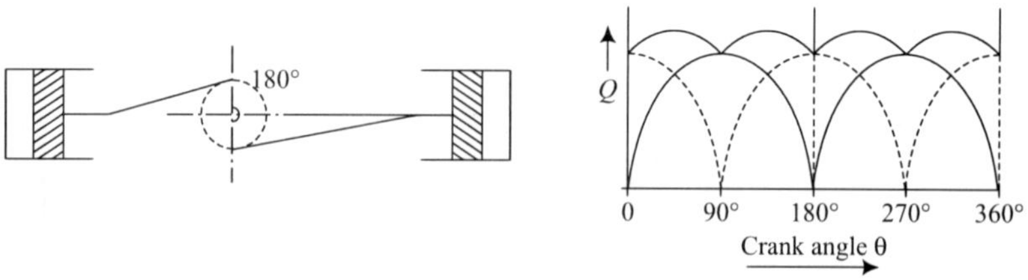

Fig. 10.5 (c) Double cylinder double acting pump

4. Three-cylinder double acting pump.

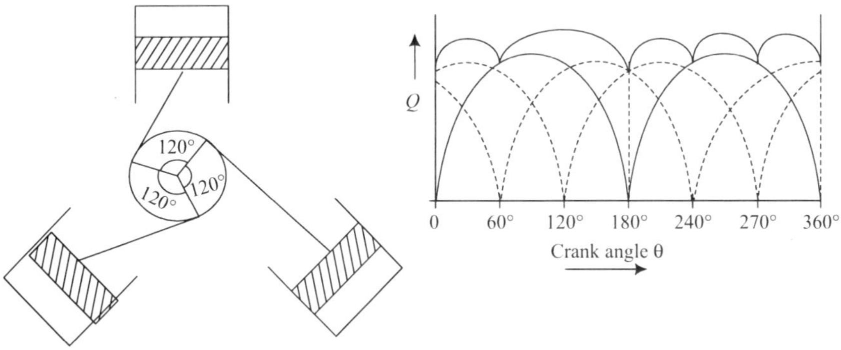

Fig. 10.5 (d) Three cylinder double acting pumps

5. Pump with air vessels.

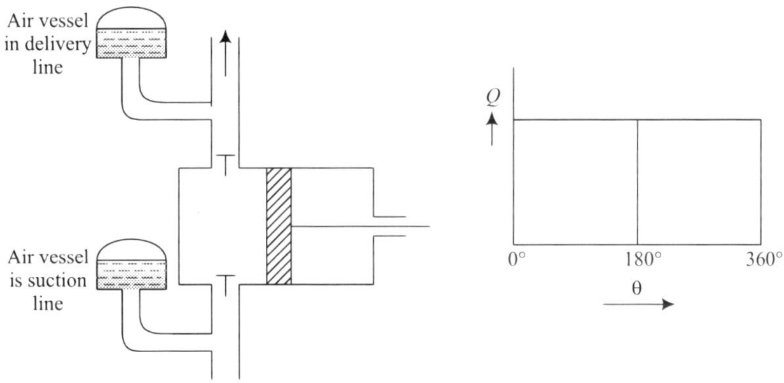

Fig. 10.5 (e) Pump with air vessels

10.4 PISTON MOTION

After the pump has been primed, the liquid follows the piston closely due to forces of cohesion and adhesion without separation. The acceleration of piston will impart head to the liquid–also called acceleration head.

The piston displacement,

$$x = r - r\cos\theta$$

where θ = crank angle

r = crank radius

Also $$\theta = \omega t$$

where ω = Angular velocity [rad/s]

t = time (seconds)

$$\therefore \quad x = r - r\cos\omega t$$

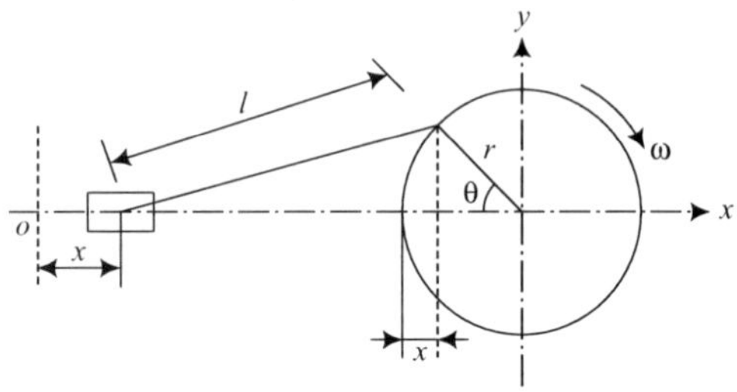

Fig. 10.6 Piston motion

The velocity of piston,

$$V_p = \frac{dx}{dt} = 0 - r(-\sin\omega t)\omega$$

$$= \omega r \sin\omega t$$

The acceleration of piston,

$$a_p = \frac{d^2x}{dt^2} = \omega^2 r\cos\omega t$$

Flow of liquid in suction pipe = Flow of liquid in the cylinder

$$Q = V_S A_S = A\omega r\sin\omega t$$

where V_S = Velocity of liquid in suction pipe (m/s)

A_S = Area of suction pipe [m^2]

A = Area of piston [m^2]

Acceleration of liquid in suction pipe,

$$a_s = \frac{A}{A_S}\omega^2 r\cos\omega t$$

$$= \frac{A}{A_S}\omega^2 r\cos\theta$$

Accelerating force = Mass × acceleration

$$= \rho A_S l_S \times \frac{A}{A_S}\,\omega^2 r\cos\theta \text{ [N]}$$

where l_S = length of suction pipe [m]

Pressure, $$p_S = \rho \frac{A}{A_S} \cdot l_S \frac{A}{A_S} \omega^2 r \cos\theta \text{ [N/m}^2\text{]}$$

Acceleration pressure head,

$$Ha_S = \frac{\rho}{\rho g} \frac{A}{A_S} l_S \frac{A}{A_S} \omega^2 r \cos\theta \text{ [m]}$$

$$Ha_S = \frac{l_S}{g} \frac{A}{A_S} \omega^2 r \cos\theta \qquad ...(1)$$

At the beginning of stroke, $\theta = 0$

$$\therefore \quad Ha_S = \frac{l_S}{g} \cdot \frac{A}{A_S} \omega^2 r$$

At the mid of stroke, $\theta = 90°$

$$Ha_S = 0$$

At the end of stroke, $\theta = 180°$

$$Ha_S = -\frac{l_S}{g} \cdot \frac{A}{A_S} \omega^2 r$$

$$\therefore \quad (Ha_S)_{\max} = \frac{l_S}{g} \frac{A}{A_S} \omega^2 r$$

Maximum acceleration head occurs when the piston is at dead centres. The length of connecting rod has been neglected, being very long.

10.5 INDICATOR DIAGRAM

The work input to the pump can be calculated from the indicator diagram which shows the pressure of liquid in the pump cylinder corresponding to crank position during suction and delivery strokes. The piston of the indicator moves up and down against the spring pressure as the cylinder pressure of the pump varies. The pencil moves with the indicator piston. The drum carrying the graph paper wrapped on it oscillates by reduction motion proportional to pump piston stroke.

The area of indicator diagram = work done per revolution for single acting pump.

(The area of indicator diagram) × 2 = work done per revolution for double acting pump

$$W.D. = \rho g Q H = \frac{\rho g A L N}{60} (H_S + H_d)$$

$$= \frac{\rho g A N}{60} [L(H_S + H_d)]$$

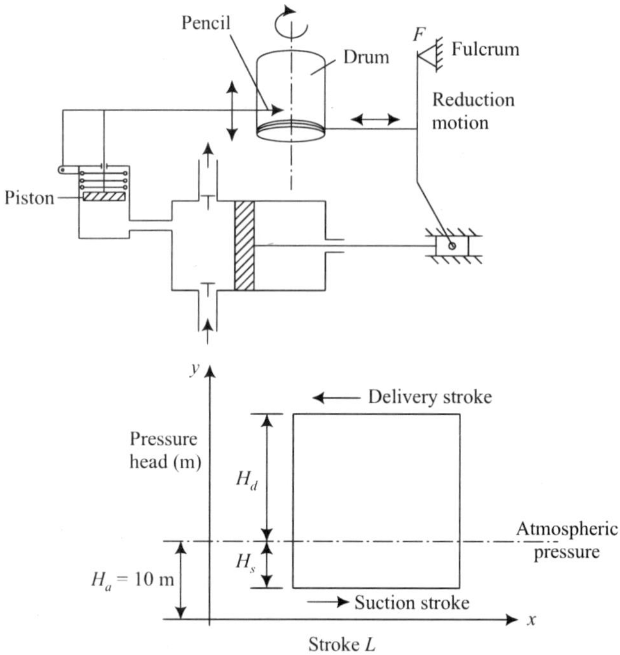

Fig. 10.7 Indicator diagram

$$W.D. = \frac{\rho g A N}{60} \times \text{area of indicator diagram.}$$

The area of indicator diagram is proportional to work done for running the pump.

The indicator diagram shown above is for ideal operating conditions. In actual operation, the performance as well as indicator diagram are affected by various factors.

10.6 FACTORS AFFECTING PERFORMANCE OF RECIPROCATING PUMPS

The actual performance of reciprocating pump is affected by the following factors:

1. Slip
2. Acceleration head of liquid in suction and delivery pipes
3. Friction losses
4. Liquid separation
5. Air vessels

10.6.1 Effect of Slip of Liquid

The actual liquid delivered by the pump is usually less than theoretical piston swept volume due to the following reasons.

1. Leakage through the valves, glands and piston rings.
2. Delay in closing of suction and delivery valves.
3. Liberation of air from the liquid being handled.

The difference in volume delivered from theoretical volume is called slip.

$$\text{Slip} = [\text{Volume swept/stroke}] - [\text{Actual discharge/stroke}]$$

The coefficient of discharge,

$$C_d = \frac{(\text{Actual discharge/stroke})}{(\text{Volume swept/stroke})}$$

$$\text{Percentage slip} = \left(\frac{Q_{th} - Q_{act}}{Q_{th}}\right) \times 100$$

$$= (1 - C_d) \times 100$$

Sometimes by adjusting the valve timing (delivery valve opens before suction stroke is completed), C_d can be more than unity. The pump would be working with *negative slip* and actual discharge will be more than theoretical discharge.

10.6.2 Losses of Head in Suction Pipe

In order to calculate losses in the suction pipe of the pump, apply Bernaulli's equation at points (1) and (2), in Fig. 10.8.

$$H_1 = H_2 + H_S + \frac{V_2^2}{2g} + Ha_S + \sum \Delta H_l$$

where H_1 = pressure at top of suction level

= 1 atm. = 10 m.

H_2 = Pressure in the cylinder which should be more than vapour pressure H_v, (~ 2.5 m at normal liquid temperature)

H_S = Suction height

$\frac{V_2^2}{2g}$ = Velocity head

$$= \frac{\left(\frac{A}{A_S}\omega r \cos\theta\right)^2}{2g}$$

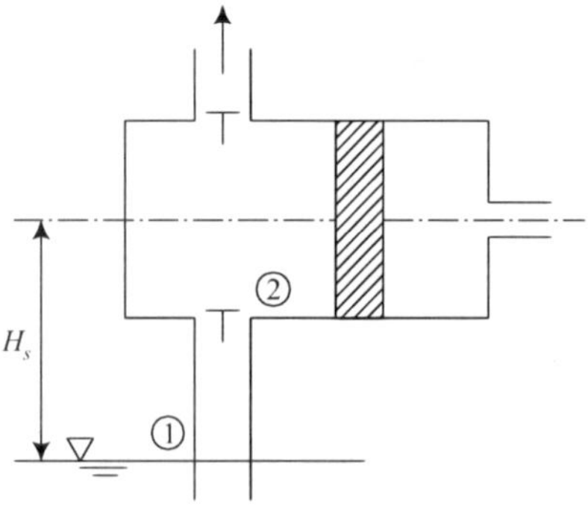

Fig. 10.8 Suction losses

Ha_S = Acceleration suction head due to inertia forces

$$= \frac{l_S}{g} \cdot \frac{A}{A_S} \omega^2 r \cos\theta$$

$$Ha_S = \frac{l_S}{g} \frac{A}{A_S} \omega^2 r,$$ i.e. the maximum head at the beginning and end of suction stroke.

$= 0$ at $\theta = 90°$, i.e. at the mid-stroke

$\sum \Delta H_l$ = sum of losses

= Pipe friction + Friction losses in the valves and other fittings + Head required to open the valves.

$$\text{Pipe friction} = \frac{4 f l_S V_S^2}{2 g d_S} = \frac{4 f l_S}{2 g d_S} \left(\frac{A}{A_S} \omega r \cos\theta \right)^2$$

$$= \frac{4 f l_S}{2 g d_S} \left(\frac{A}{A_S} \omega r \right)^2$$ at $\theta = 90°$, i.e., maximum

= 0 at beginning and end of suction stroke.

The indicator diagram including all losses and acceleration head is shown in Fig. 10.9.

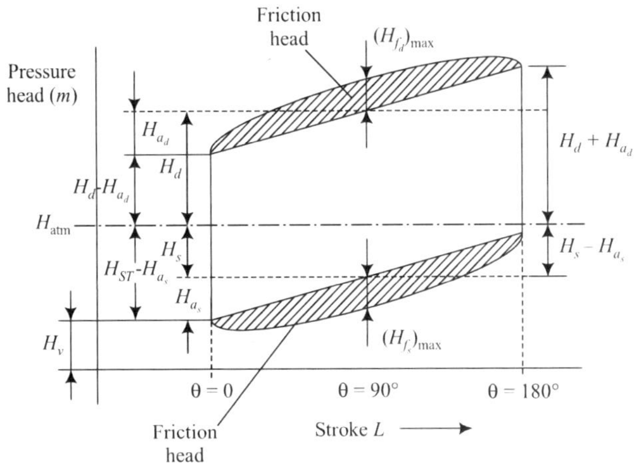

Fig. 10.9 Effect of friction and acceleration on indicator diagram

The pressures in the cylinder are function of cos θ and are tabulated below.

Table 10.1 Pressures in the cylinder

Sl. No.	*Stroke length*	*Suction stroke*	*Delivery stroke*
1.	Beginning, θ = 0°	$H_{atm} - (H_S + H_{a_S})$ $H_{f_S} = 0$	$H_{atm} + (H_d + H_{a_d})$ $H_{f_d} = 0$
2.	Middle, θ = 90°	$H_{atm} - (H_S + H_{f_S})$ $H_{a_S} = 0$	$H_{atm} + (H_d + H_{f_d})$ $H_{a_d} = 0$
3.	End, θ = 180°	$H_{atm} - (H_S + H_{a_S})$ $H_{f_S} = 0$	$H_{atm} + (H_d + H_{a_d})$ $H_{f_d} = 0$

10.6.3 Liquid Separation

Whenever liquid pressure falls below its vapour pressure, liquid boiling and separation take place causing cavitation.

Refer Fig. 10.8

$$H_2 \geq H_v$$

where H_v = Vapour pressure

$$H_2 = H_1 - H_S - \frac{V_2^2}{2g} - H_{a_S} - \sum \Delta H_l$$

H_1 is a function of altitude which is fixed for a site.

H_2 can be increased by the following methods.

(i) Reducing H_S, i.e., low suction height

(ii) Reducing $\frac{V_2^2}{2g}$, i.e., large suction pipe diameter and small piston speed.

(iii) Reducing Ha_S, i.e., small suction pipe length and small piston speed.

(iv) Reducing $\Sigma \Delta H_l$, i.e., small suction pipe length and minimum number of bends and pipe fittings.

1. Maximum pump speed

The reciprocating pumps are designed for low speed and are, therefore, driven through reduction gear or belt drive by electric motor and not directly coupled.

The maximum acceleration head,

$$(H_{a_S})_{\max} = \frac{l_S}{g} \cdot \frac{A}{A_S} \omega^2 r$$

The pressure head at the beginning of suction stroke

$$= H_{\text{atm}} - H_S - (H_{a_S})_{\max}$$

$$\geq H_v$$

$$\therefore \quad H_S + (H_{a_S})_{\max} \leq (H_{\text{atm}} - H_v)$$

$$H_{\text{atm}} = 10.3 \text{ m}$$

$$\therefore \quad 10.3 - (H_S + H_{a_S}) \leq 2.5$$

$$\text{or} \quad (H_S + H_{a_S}) \leq 7.8$$

H_S is constant for a pump installation.

$$H_{a_S} = f\left[l_S, \left(\frac{A}{A_S}\right)^2, \omega^2 r\right]$$

l_S, A_S, A and r are constants for a pump installation

$$\therefore \quad H_{a_S} = f(\omega^2)$$

$$H_{a_S} = f(N)$$

N should be small for low value of H_{a_S}.

2. Pipe layout

The friction losses depend upon the piping layout. The pipes should be laid straight without bends.

3. Air vessels

Air vessel is a closed chamber fitted on suction and delivery pipes, near the pump cylinder, to reduce acceleration head (inertia) of liquid column.

The air vessel on *suction pipe* helps:

(i) to reduce possibility of separation

(ii) to operate on high possible speeds

(iii) to increase length of suction pipe

The air vessel on *delivery pipe* helps:

(i) saving of power due to less inertia head

(ii) to ensure constant flow rate

4. Direct acting steam pump

Reciprocating pumps are normally driven by electric motors through reduction gears or *v*-belt drive. In boiler houses of process industries where steam is available the boiler feed pumps may be directly connected to a steam engine. Direct acting steam pump is more economical and reliable than electric motor driven pump. The crankshaft, flywheel, etc. are completely eliminated. The pump duty is defined as:

$$\text{Pump duty} = \frac{\rho QgH}{\dot{m}_S\, h_S \times 10^3} \text{ [kJ/kJ]}$$

where $\dot{m}_S$ = steam flow rate [kg/s]

h_S = enthalpy of steam [kJ/kg]

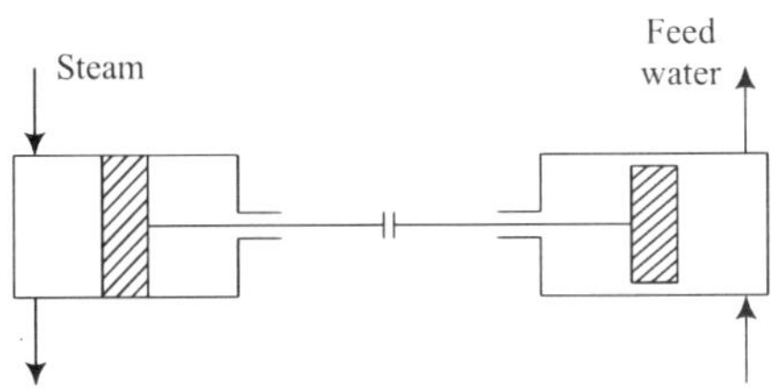

Fig. 10.10 Direct driven pump

10.6.4 Effect of Air Vessels

It has been found that use of air vessels along with reciprocating pumps ensures constant flow rate and reduces the possibility of liquid separation.

1. Air vessel on suction side

When air vessel is used, suction takes place in two steps:

(i) Liquid flows through suction pipe into air vessel.

(ii) Liquid is raised from air vessel to cylinder.

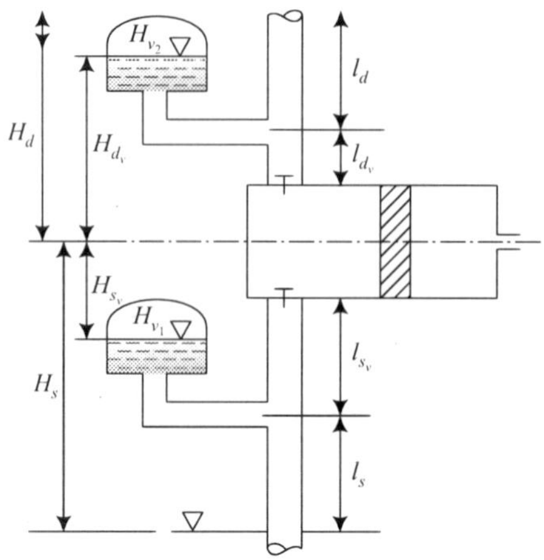

Fig. 10.11 Pump with air vessels

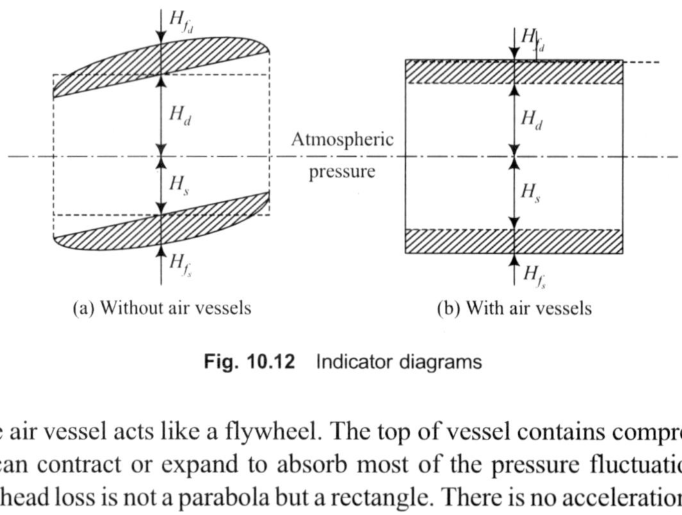

Fig. 10.12 Indicator diagrams

The air vessel acts like a flywheel. The top of vessel contains compressed air which can contract or expand to absorb most of the pressure fluctuations. The friction head loss is not a parabola but a rectangle. There is no acceleration head in the pipe except in short pipe length l_{Sv}.

2. Air vessel on discharge side

During delivery strokes, suction valve is closed and flow takes place from air vessel on delivery side which was filled in the previous stroke. Here also the friction head loss is constant and acceleration head takes place in short pipe length, l_{dv}.

When air vessels of sufficient capacity are installed on the suction and delivery pipes, the flow velocity in suction and delivery pipe can be assumed to be uniform and equal to mean flow velocity.

The discharge,

$$Q = \text{Piston area} \times \text{stroke length} \times \text{speed}$$

$$= A \times L \times \frac{N}{60}$$

$$= A \times 2r\,\frac{\omega}{2\pi} = Ar\,\frac{\omega}{\pi}$$

The flow velocity in pipes,

$$V = \frac{Q}{A_S} = \frac{A}{A_S}\frac{\omega r}{\pi}$$

Loss of head due to friction

$$Hf_S = \frac{4fl_S v^2}{2gd_S} = \frac{4fl_S}{2gd_S}\left[\frac{A}{A_S}\frac{\omega r}{\pi}\right)^2$$

Area of indicator diagram = base × height,

$$= L \times \frac{4f(l_s + l_d)}{2gd_S}\left[\frac{A}{A_S}\frac{\omega r}{\pi}\right]^2$$

Power expanded,

$$P_i = \frac{\rho g A N}{60} \times L\,\frac{4f(l_s + l_d)}{2gd_S}\left[\frac{A}{A_S}\frac{\omega r}{\pi}\right]^2$$

Power expanded in pumps without air vessel,

$$P_i' = \frac{\rho g A N}{60} \times \frac{2}{3} \times L \times \frac{4f(l_s + l_d)}{2gd_S}\left[\frac{A}{A_S}\omega r\right]^2$$

Therefore,
$$\frac{P_i}{P_i'} = \frac{\dfrac{\rho g A N}{60} \times \dfrac{4f(l_S + l_d)}{2gd_S}\left[\dfrac{A}{A_S}\dfrac{\omega r}{\pi}\right]^2}{\dfrac{\rho g A N}{60} \times \dfrac{2}{3} \times \dfrac{4f(l_S + l_d)}{2gd_S}\left[\dfrac{A}{A_S}\omega r\right]^2}$$

$$= \frac{3}{2\pi^2}$$

Saving in power expanded due to air vessels,

$$= \frac{P_i' - P_i}{P_i'} = 1 - \frac{P_i}{P_i'} = \left(1 - \frac{3}{2\pi^2}\right) = 0.848 = 84.8\%$$

QUESTION BANK NO. 10

1. Explain the working of a reciprocating pump.
2. With the help of a suitable layout diagram, explain the performance of a reciprocating pump.
3. How is the pulsating flow of a reciprocating pump converted into continuous flow? Explain the different methods.
4. Explain the formation of acceleration head with the help of piston motion.
5. Explain the function of air vessels in a reciprocating pump.
6. What is an indicator diagram of a reciprocating pump? Sketch the theoretical indicator diagram for a single acting reciprocating pump not fitted with air vessels. With the help of this diagram, explain clearly the effect of acceleration and friction on both suction and delivery strokes.
7. Explain how the indicator diagram is modified if air vessels are provided on both suction and delivery pipes.
8. What is liquid separation? How it can be controlled in reciprocating pumps?
9. Show that the maximum acceleration head in a reciprocating pump without air vessel is given by:

$$H_a = \frac{l}{g} \times \frac{A}{A_S}\, \omega^2 r \text{ with usual notations}$$

TUTORIAL SHEET NO. 10

1. A single acting reciprocating pump has a piston diameter 15 cm and stroke length 30 cm. The centre of the pump is 5 m above the level of water and 33 m below delivery water level. The length of suction and delivery lines are 6.5 m and 39 m respectively and both pipes have same diameter of 7.5 cm. If the pump is working at 30 rpm, find the pressure head on the piston at the beginning, middle and end of both suction and delivery strokes, also find the power required to drive the pump. Assume atmospheric pressure as 10.3 m of water and friction factor of 0.01.

   Solution The piston area,

$$A = \frac{\pi}{4}\left(\frac{15}{100}\right)^2 = 0.0176 \text{ m}^2$$

$$L = 0.30 \text{ m}$$
$$H_S = 5 \text{ m}$$
$$H_d = 33 \text{ m}$$
$$l_S = 6.5 \text{ m}$$
$$l_d = 39 \text{ m}$$

The area of suction and delivery pipes,

$$A_S = \frac{\pi}{4}(0.075)^2 = 4.4 \times 10^{-3} \text{ m}^2$$
$$N = 30 \text{ rpm}$$
$$\omega = \frac{2\pi N}{60} = \frac{2\pi \times 30}{60} = \pi \text{ rad/s.}$$

The crank radius, $r = \dfrac{L}{2} = 0.15$ m

$$H_{atm} = 10.3 \text{ m}$$
$$Ha_S = \frac{l_S}{g}\frac{A}{A_S}\ \omega^2 r \cos\theta$$
$$(Ha_S)_{max} = \frac{l_S}{g}\cdot\frac{A}{A_S}\ \omega^2 r$$
$$= \frac{6.5}{9.81} \times \frac{0.01767}{4.4 \times 10^{-3}}\ \pi^2 \times 0.15 = 3.94 \text{ m.}$$
$$(Hf_S)_{max} = \frac{4 f l_S}{2 g d_S}\left(\frac{A}{A_S}\omega r\right)^2$$
$$= \frac{4 \times 0.01 \times 6.5}{2 \times 9.81 \times 0.075}\left[\frac{0.01767}{4.4 \times 10^{-3}} \times \pi \times 0.15\right]^2$$
$$(Hf_S)_{max} = 0.63 \text{ m}$$
$$H_d = 33 \text{ m}$$
$$(Ha_d)_{max} = \frac{l_d}{g}\cdot\frac{A}{A_d}\ \omega^2 r = \frac{39}{9.81}\left(\frac{0.01767}{4.4 \times 10^{-3}}\right)\pi^2 \times 0.15$$
$$= 23.64 \text{ m}$$

$$(Hf_d)_{max} = \frac{4fl_d}{2gd_d}\left(\frac{A}{A_d}\omega r\right)^2 = \frac{4 \times 0.01 \times 39}{2 \times 9.81 \times 0.075}\left[\frac{0.01767}{4.4 \times 10^{-3}} \times \pi \times 0.15\right]^2$$

$$= 3.78 \text{ m}$$

Stroke length	*Suction stroke*	*Delivery stroke*
1. Beginning, $\theta = 0$	$H_{atm} - (H_S + H_{a_S})$ $= 10.30 - (5 + 3.94)$ $= 1.36$ m	$H_{atm} + (H_d + H_{a_d})$ $= 10.3 + 33 + 23.64$ $= 66.94$ m
2. Middle, $\theta = 90°$	$H_{atm} - (H_S + H_{f_S})$ $= 10.3 - (5 + 0.63)$ $= 4.67$ m	$H_{atm} + (H_d + H_{f_d})$ $= 10.3 + 33 + 3.78$ $= 47.08$ m
3. End, $\theta = 180°$	$H_{atm} - (H_S - H_{a_S})$ $= 10.3 - (5 - 3.94)$ $= 9.24$ m	$H_{atm} + (H_d - H_{a_d})$ $= 10.3 + 33 - 23.64$ $= 19.66$ m

Total head delivered.

$$H = \text{Static head} + \text{friction head losses} + \text{acceleration heads}$$

$$= (H_S + H_d) + (H_{f_S} + H_{f_d}) + (H_{a_S} + H_{a_d})$$

$$= 5 + 33 + (0.63 + 3.78) + (3.94 + 23.64)$$

$$= 69.99 \text{ m}$$

The power required to drive the pump,

$$P_i = \rho QgH = \frac{\rho g}{1000}\frac{ALN}{60}H$$

$$= \frac{9810}{1000} \times 0.0176 \times 0.3 \times \frac{30}{60} \times 69.99 = \mathbf{1.82\ kW} \quad \text{Ans}$$

2. A single acting reciprocating pump runs at 30 rpm. The diameter of plunger is 15 cm and crank radius is 15 cm. The suction pipe is 10 cm in diameter and 9 m long. Calculate the maximum permissible value of suction lift H_S, if the separation takes place at 2.6 m of water. Take atmospheric pressure head as 10.3 m of water.

   Solution The given data:

$$N = 30 \text{ rpm}; \ \omega = \frac{2\pi N}{60} = \frac{2\pi \times 30}{60} = \pi \text{ rad/s}$$

$$A = \frac{\pi}{4}(0.15)^2 = 0.01767 \text{ m}^2$$

$$r = 0.15 \text{ m}$$

$$d_s = 0.1 \text{ m} \quad A_S = \frac{\pi}{4}(0.1)^2 = 7.85 \times 10^{-3} \text{ m}^2$$

$$l_S = 9 \text{ m}$$

$$H_v = 2.6 \text{ m}$$

$$H_{atm} = 10.3 \text{ m}$$

$$H_{atm} = H_v + H_S + \frac{V_2^2}{2g} + H_{a_S} + \sum \Delta H_l.$$

Neglect $\frac{V_2^2}{2g}$ and valve losses

$$\therefore \quad H_{atm} = H_v + H_S + H_{a_S} + H_{f_S}$$

$$= 2.6 + H_S + \frac{l_S}{g}\omega^2 r \times \frac{A}{A_S} \cos\theta + \frac{2fl_S}{2gd_s}\left(\frac{A}{A_S}\omega r \sin\theta\right)^2$$

At $\theta = 0$, i.e., beginning of strokes

$$10.3 = 2.6 + H_S + \frac{9}{9.81}\pi^2 \times 0.15 \times \frac{0.01767}{7.85 \times 10^{-3}} + 0$$

$$= 2.6 + H_S + 3$$

$$\therefore \quad H_S = \mathbf{4.7\ m} \quad \text{Ans}$$

3. A single acting reciprocating pump has a plunger of 18 cm diameter and a stroke of 20 cm and supplies water from a sump 4 m below the pump through a pipe 6.5 m long and 5 cm in diameter. It delivers water to a reservoir 15 m above the pump through a pipe 3.5 cm in diameter and 16 cm long. If separation occurs at 8 cm of water, find the maximum speed at which the pump may be operated without separation.

Solution The given data:

$$A = \frac{\pi}{4} \times 0.18^2 = 0.02545 \text{ m}^2$$

$$l_S = 6.5 \text{ m}$$

$$d_s = 0.05 \text{ m}$$

$$A_S = \frac{\pi}{4} \times (0.05)^2 = 1.963 \times 10^{-3} \text{ m}^2$$

$$H_S = 4 \text{ m}$$
$$H_d = 15 \text{ m}$$
$$d_d = 0.35 \text{ m}$$
$$l_d = 16 \text{ m}$$
$$H_v = 0.08 \text{ m}$$

At the beginning of suction stroke

$H_{atm} = H_S + (H_a)_{max} + H_v$ (Neglecting friction head and velocity head)

$$= H_S + \frac{l_S}{g} \cdot \frac{A}{A_S} \omega^2 r + H_v$$

$$10.3 = 4 + \frac{6.5}{9.81} \times \left(\frac{0.02545}{1.963 \times 10^{-3}}\right)\left(\frac{2\pi N}{60}\right)^2 \cdot \frac{0.2}{2} + 0.08$$

$$10.3 = 4.08 + 0.009417 N^2$$

$\therefore \quad N = 25.7$ rpm

At the end of delivery stroke,

$$H_{atm} = H_{ad} - H_d + H_v$$

$$10.3 = \frac{16}{9.81}\left(\frac{0.18}{0.35}\right)^2 \left(\frac{2\pi N}{60}\right)^2 + \frac{0.2}{2} - 15 + 0.08$$

$$10.3 = 0.0004 N^2 - 15 + 0.08$$

$\therefore \quad$ N = 251 rpm

$\therefore \quad$ Maximum speed = **25.7 rpm** Ans.

4. The bore diameter of a double acting twin cylinder pump is 10 cm and the length of stroke is 15 cm. When running at 60 rpm, it discharges 4.5 *l/s* and has a total lift of 10 m. What is the percentage of slip? If the required liquid pumped has a specific gravity of 0.96, what is the power required assuming efficiency of the unit to be 60%?

Solution

$$A = \frac{\pi}{4} (0.1)^2 = 7.854 \times 10^{-3} \text{ m}^2$$

$$L = 0.15 \text{ m}$$

$$N = 60 \text{ rpm}; \ \omega = \frac{2\pi N}{60} = 2\pi \text{ rad/s}$$

$$Q = 4.5\ l/s = 4.5 \times 10^{-3}\ \text{m}^3/\text{s}$$

$$H = 10\ \text{m}$$

$$\text{Slip} = (\text{Volume swept/stroke}) - (\text{Actual discharge/stroke})$$

$$= \frac{2\,ALN}{60} \times 2 - 4.5 \times 10^{-3}$$

$$= 2(7.854 \times 10^{-3}) \times 0.15 \times \frac{60}{60} \times 2 - 4.5 \times 10^{-3}$$

$$= 0.0471 - 0.045 = 0.0021\ \text{m}^3/\text{s}$$

$$\%\ \text{age slip} = \frac{0.0021}{0.0471} \times 10^2 = \mathbf{4.45\%} \quad \text{Ans.}$$

Power required,

$$P_i = \frac{\rho Q g H}{1000 \times \eta} = \frac{960 \times 4.5 \times 10^{-3} \times 9.81 \times 10}{1000 \times 0.6}$$

$$= \mathbf{0.70\ kW} \quad \text{Ans}$$

5. A single acting reciprocating pump operating at 120 rpm has a piston diameter of 200 mm and stroke of 300 mm. The suction and delivery heads are 4 m and 20 m. If the efficiency of both suction and delivery strokes is 75 per cent, determine the power required by the pump (6 kW).

CHAPTER 11

Special Pumps

11.1 INTRODUCTION

The centrifugal pumps and reciprocating pumps are most widely used pumps. But there are special pumps designed for specific applications. The following pumps have been selected for discussions.

1. *Axial flow pumps*

   These pumps include vertical turbine pump, deep well pump, submersible pump, borehole pump, etc.

2. *Rotary displacement pump*

   These pumps include gear pump, lobe pump, vane pump, screw pump. liquid ring pump.

3. *Jet pumps*

   These pumps include air lift pump and water jet pump.

4. *Fluid coupling*

   These include fluid coupling and torque converter.

11.2 AXIAL FLOW PUMP

Axial flow pumps or propeller pump work as converse of a propeller turbine. Pressure is developed by flow of liquid over blades of airfoil section. If Kaplan turbine runner with adjustable blades is used, the pump may be called Kaplan pump.

These pumps are designed for large capacities and low heads below 1.2 m. These pumps have high specific speeds. For the sake of compactness, these pumps are usually designed with vertical shaft disposition. These are used for land drainage, flood control, storm water disposal, irrigation, condenser circulation pump in power plants and other low head pumping jobs.

1. Design constants

The design procedure is similar to that for propeller/Kaplan turbine. The pump constants are:

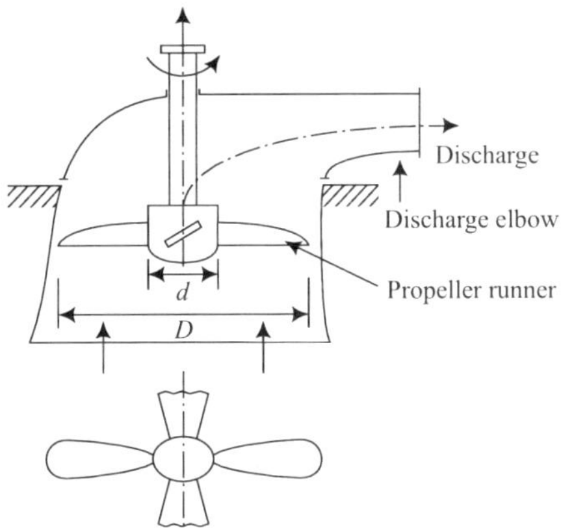

Fig. 11.1 Axial pump

Speed ratio, $\phi = \dfrac{V_b}{\sqrt{2gh}} = 2 \text{ to } 3$

Flow ratio, $\psi = \dfrac{V_f}{\sqrt{2gH}} = 0.4 \text{ to } 0.8$

Diameter ratio, $\dfrac{d}{D} = 0.4 \text{ to } 0.8$

Hydraulic efficiency, $\eta_h = 0.75 \text{ to } 0.92$

Volumetric efficiency, $\eta_{vol} = 1$

Mechanical efficiency, $\eta_m = 0.94 \text{ to } 0.98$

Overall efficiency, $\eta_{overall} = \eta_h \times \eta_m = 0.70 \text{ to } 0.90$

The maximum lift ≤ 1 m

The discharge, $Q = \dfrac{\pi}{4}(D^2 - d^2)V_f$

The driving power, $P_i = \dfrac{\rho Q g H}{1000 \times \eta_{overall}}$ [kW]

2. Characteristic curves

The characteristic curves of propeller pumps are shown in Fig. 11.2.

(i) *Q-H curve* is saddle shaped, which is obtained due to vane lift being reached at low capacities and increased angle of attack. Low pressure pumps have constantly drooping characteristic.

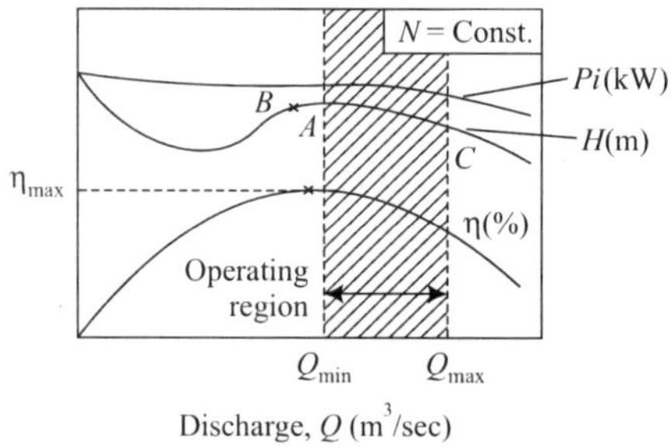

Fig. 11.2 Characteristic curves

(ii) *Power curve* is approximately horizontal initially and then drops with increase of flow. Axial pumps are started on load with delivery valve open.

(iii) *Efficiency curve* has a pronounced peak. The efficiency changes abruptly when operated away from optimum design conditions.

3. Operating region

The operating region is selected within the steady portion of the characteristic curve ($A - C$) to the right of peak B.

The minimum discharge, Q_{min} corresponds to $(H_A)_{\text{max}} = 0.9\ H_B$.

The maximum discharge, Q_{max} corresponds to $\eta_c = 0.9\ \eta_{\text{max}}$.

4. Capacity control

The capacity control of axial pumps can be affected by:

(i) Speed regulation, i.e., most efficient

(ii) Throttling is the most unfavourable economical operation as power grows as capacity is reduced

(iii) Adjustable inlet guide vanes where driving power can be used economically

(iv) Adjustable impeller vane is also power efficient control.

11.3 DESIGN OF AXIAL FLOW PUMP

In an axial flow pump, the energy is carried from the shaft to the flow by an impeller consisting of cantilever vanes (2 to 6 in number) fixed on the hub. The performance of axial flow pump is considered using the theory of cascades of profiles.

In view of the very low head produced by an axial flow impeller, the skin friction loss or drag acquires a greater importance than in centrifugal or mixed flow impellers. For that reason, a high degree of impeller vane streamlining and polishing is required to obtain the optimum peak efficiency. To satisfy this condition and that of mechanical strength, the impeller vane profiles take the form of airfoils. Although developed primarily for the airplane supporting wing application, airfoils have found wide use in the field of axial flow pumps and turbo machines.

A great many airfoil sections have been tested and classified according to their curvature and thickness.

11.3.1 Airfoil Section

Airfoils are considered as made up of a certain profile thickness form disposed about certain mean line as shown in Fig. 11.3. The form of the mean line determines completely most of the important hydraulic properties of airfoils, whereas thickness is dictated by the strength requirements. All good airfoils have nearly the same thickness variation along the mean line, the maximum thickness being different for different profiles. The maximum distance from the chord to the mean line is called camber (C) and is usually expressed in percentage of the chord length or C/l.

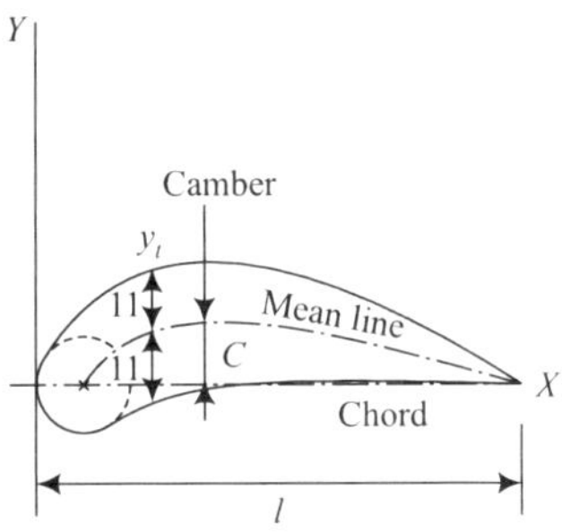

Fig. 11.3 Airfoil section

11.3.2 Definitions

Each N.A.C.A. profile is designated by a four digit number.

Example. 4312.

where 4 = Camber of the mean line

3 = Location of the camber four leading edge in lengths of chord lengths.

12 = Maximum vane thickness in percentage of the chord (l).

1. Angle of attack (α)

It is an acute angle between the vane chord and direction of relative velocity of the flow (V_{r_a}).

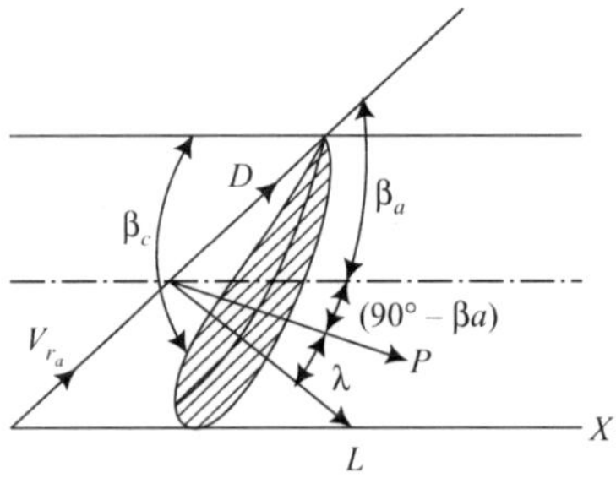

Fig. 11.4 Forces acting on airfoil

2. Aspect ratio

It is the ratio of the length of airfoil and the length of the chord. In a pump, liquid is confined between the hub and the casing sidewalls and is not free to escape radially, an aspect ratio of infinity is used for axial flow pumps.

3. Lift (L)

It is the component of the force acting on the airfoil normal to the direction of the approaching undisturbed flow.

$$L = C_L bl\rho \frac{V_{r_a}^2}{2}$$

where C_L = experimental coefficient of lift

b = width of airfoil

l = length of chord of airfoil

ρ = density of fluid

V_{r_a} = undisturbed relative liquid velocity

4. Drag (D)

It is the component of the force acting on the airfoil in the direction of flow.

$$D = C_D bl\rho \frac{V_{r_a}^2}{2}$$

where C_D = experimental coefficient of drag.

C_L and C_D depend upon the profile of airfoil, angle of attack α and aspect ratio. C_D is very small as compared to C_L. These values have been experimentally determined for a large number of profiles.

5. Relation between lift and drag

The airfoil efficiency,

$$\eta_{\text{airfoil}} = \frac{L}{D}$$

The gliding angle,

$$\lambda = \frac{C_D}{C_L}$$

The airfoil efficiency is more for thin airfoils. The maximum value of C_L is obtained with 12% airfoil thickness.

If C_L stands for head, α standards for capacity and L/D stands for efficiency, then the curves between C_L and α resemble the ordinary head-capacity curve of an axial flow pump.

11.3.3 Airfoil Theory

The airfoil theory of axial flow impeller design establishes a connection between the lift coefficient of airfoil test data and impeller total head.

The design procedure consists of:

1. Selection of the section of suitable airfoil properties for several radii of impeller.
2. Determination of the vane setting on the hub or chord angle β_c for each section.
3. Assume constant head for several streamlines.
4. Establish vane curvature based on airfoil experiment knowledge.

Impeller design is entirely experimental.

11.3.4 Limitations of Airfoil Theory

The airfoil theory fails to help in the selection of the following design elements of the impeller and are selected on the basis of previous experience only.

1. Hub ratio
2. l/t ratio at each radius
3. Speed v/s specific speed.
4. Axial velocity
5. Impeller diameter.

Presently airfoil theory has become obsolete and the method of rotating profiles in suitable casing is used for the design of axial flow pumps. New design methods based on inlet and outlet velocity triangles have been evolved for the design of axial flow impeller.

11.3.5 Total Head Equation

If a cylindrical cut is made through an axial flow impeller and the cylinder is developed onto plane, a row of vane profiles will result. The action of liquid on the profile can be considered similar to that taking place on an airfoil in a wind tunnel, provided the relative velocity V_{r_a} is an average velocity of approach and discharge V_{a_1} and V_{a_2} before and after the vane at a distance whose effect of the flow through the row of vanes is equalized.

The effect of the cascade arrangement of vanes on C_L is neglected.

The total force (P) exerted by the liquid per unit wavelength is the resultant of lift (L) and drag (D) and makes an angle λ with the direction of lift.

$$P = \sqrt{L^2 + D^2}$$

$$\tan \lambda = \frac{D}{L}$$

The total force P forms an angle $[90° - (\beta_a + \lambda)]$ with the direction of peripheral velocity (V_b) of the impeller.

The tangential component, $P_w = P \cos [90° - (\beta_a + \lambda)]$

$$\therefore \quad P_w = P \sin (\beta_a + \lambda)$$

Work done per second,

$$W.D. = V_b P \sin (\beta_a + \lambda)$$

Consider a slice of the vane between two consecutive cylindrical surfaces of radius r and $r + dr$, the work per second for Z vanes is (Fig. 11.5)

$$Z(W.D.) = ZP\ dr\ V_b \sin (\beta_a + \lambda)$$

If dQ is the volume of liquid included between the two cylindrical surfaces, the work per kg of liquid is

$$W.D./\text{kg} = \frac{Z(W.D.)}{dQ\, r} = \frac{ZP\,dr\,V_b \sin (\beta_a + \lambda)}{dQr}$$

$$dQ = Zt\ dr\ V_f$$

$$P = \frac{L}{\cos \lambda}$$

$$L = \frac{C_L\, r\, V_{r_a}^2 \cdot A}{2g}$$

where $A = l \times 1 = l$ = area per unit length

$$\therefore \quad P = \frac{C_L\, rl\, V_{r_a}^2}{2g \cos \lambda}$$

$$W.D./\text{kg} = \frac{C_L\ lr\ V_{r_a}^2 \sin(\beta_a + \lambda)}{t \cdot V_f\, 2g \cos\lambda}$$

$$C_L \frac{l}{t} = (W.D./\text{kg})\,\frac{2g}{V_{r_a}^2} \cdot \frac{V_f}{V_b} \cdot \frac{\cos\lambda}{\sin(\beta_a + \lambda)} \qquad \text{...(1)}$$

The theoretical head,

$$H_t = \frac{H}{\eta_h}$$

where η_h = Hydraulic efficiency

V_{r_a} and β_a are obtained from velocity diagram.

where $V_{b_1} = V_{b_2}$

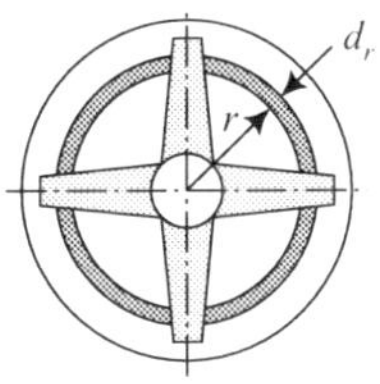

Fig. 11.5 Axial flow pump section

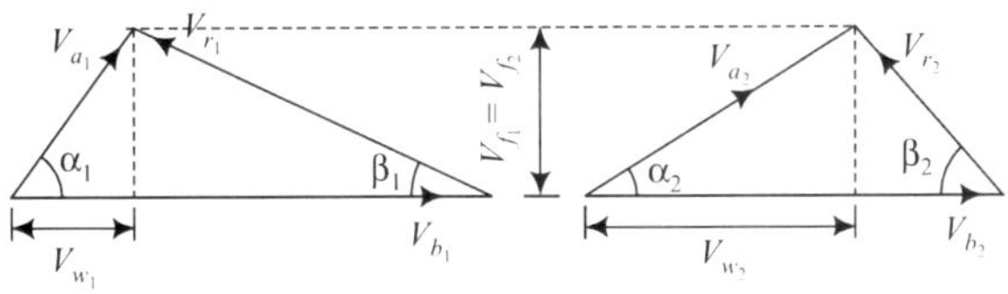

Fig. 11.6 Velocity diagrams

The average tangential component $= \dfrac{V_{w_2} - V_{w_1}}{2}$

$$V_{w_2} - V_{w_1} = \frac{gH_t}{\eta_h}$$

Assume $V_{w_1} = 0$ and V_{w_2} is obtained from velocity triangle.

From equation (1), the value of $C_L \dfrac{l}{t}$ will be low for high values of V_b and V_{r_a}. The large radii l/t and C_L will be lower than those values at the hub.

For higher heads, higher values of C_L and l/t are required, i.e., more vane area and a higher camber of profile.

The values of $C_L\left(\frac{l}{t}\right)$ are calculated for different values of impeller radii using equation (1), using constant values of H_t and V_f along radius. Experience is necessary to select proper values of l/t to obtain C_L and vane profile in order to produce an efficient impeller and head-capacity and efficiency curves of good shape.

Lift and Head Coefficients

Let $\cos\lambda = 1 \; (\because \lambda \approx 1°)$

$$\frac{V_f}{V_r} = \sin\beta_a$$

$$H_t = \frac{V_{b_2}V_{w_2}}{g}$$

Substitute these values in equation (1).

$$\therefore \quad C_L\frac{l}{t} = \frac{2V_{w_2}}{V_{r_2}} = \frac{2(\beta_2 - \beta_1)}{\sin\beta_a}$$

11.4 VERTICAL TURBINE PUMP

It is a multistage pump installed in vertical position to save space. The type of impeller, i.e., closed, semi-closed or open, depends upon the properties of the liquid being pumped. The impellers can be centrifugal (radial), mixed flow or propeller type depending upon the head and discharge requirements and the specific speed of the pump. The impellers work as reverse of reaction turbine impeller.

The main components are shown in Fig. 11.7. The liquid is sucked through a strainer and foot valve and delivered through vertical column pipe. The bearing can be oil lubricated or water lubricated.

The vertical turbine pumps are used for pumping water from the sea, river or deep lake. These are also used to pump underground water as deepwell or borewell pumps. In order to overcome the difficulty of limitation of suction lift due to NPSH requirements, the pump and suction pipe are immersed in water.

A submersible pump is also a vertical turbine pump where the electric motor is lowered along with pump inside the column pipe. The motor is designed water proof and avoids the need of long line shaft.

The hydraulic performance is similar to multi-stage centrifugal pumps.

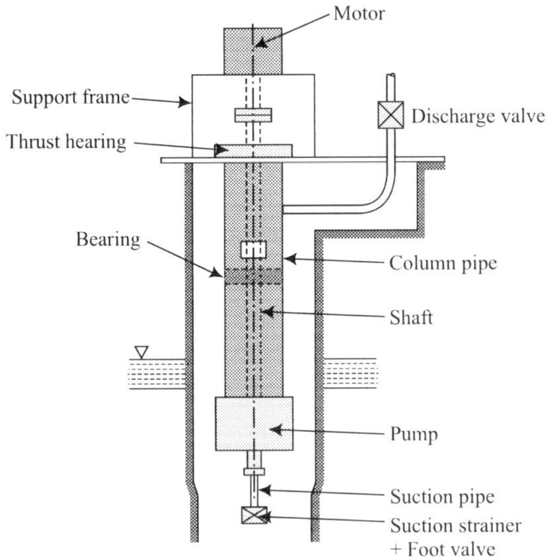

Fig. 11.7 Vertical turbine pump

11.5 ROTARY DISPLACEMENT PUMPS

The centrifugal pumps give uniform discharge and reciprocating pumps give high pressure and positive displacement of liquid. The rotary positive displacement pumps have more or less both the advantages and are used for small flows and relatively high pressures. These pumps are primarily used as power sources in the hydraulic control systems and oil lubrication systems for turbines, motors and machine tools.

The pumping elements one or more in number are directly coupled to the driver and rotate in a stationary casing. These pumps are very compact and have the self-priming ability. However, these pumps are slow speed machines. There are many designs of rotary pumps. However, gear pump, lobe pump, sliding vane pump and screw pump are popular designs.

The operating characteristics of main rotary pumps are compared in Table 11.1.

Table 11.1 Operating characteristics of rotary pumps

S. No.	*Type of pump*	*Max. capacity*	*Max. pressure*	*Max. speed*
1.	Gear pump	200 l/min.	5 MPa	1000 rpm.
2.	Screw pump	300 m^3/hr.	20 MPa	8000 rpm.
3.	Sliding vane pump	500 l/min.	20 MPa	500-1200 rpm.
4.	Lobe pump	100 l/min.	10 MPa	1000 rpm.

11.5.1 Gear Pump

The pump consists of two identical intermeshing spur gears rotating in a casing with small clearance. The drive gear is connected to the driver whose shaft extends into the casing through a stuffing box. The other gear is idle.

The gears rotate in the direction as shown. The liquid is sucked, trapped between the gear teeth and discharged.

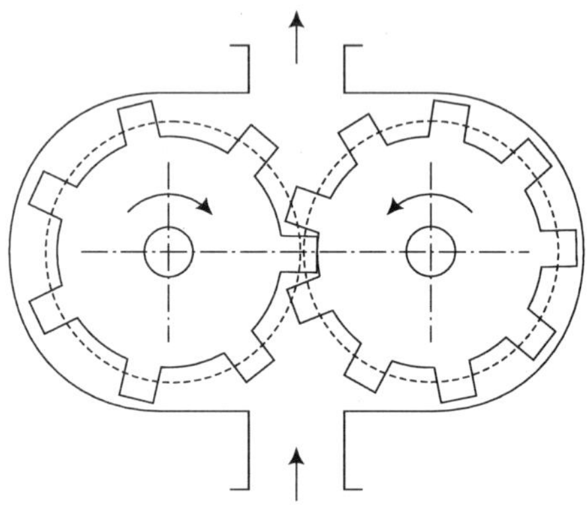

Fig. 11.8 Gear pump

The theoretical volume displaced by the pump per second,

$$Q_{\text{th}} = (2\,LAn)\frac{N}{60}\ [\text{m}^3/\text{s}]$$

where L = Axial length of teeth in each gear wheel [m]

A = Area enclosed between two adjacent teeth and casing [m^2]

n = Number of teeth in each gear wheel.

N = Speed [rpm]

Actual discharge,

$$Q_a = Q_{\text{th}} \times \eta_v$$

where η_v = volumetric efficiency [0.7 to 0.9]

11.5.2 Lobe Pump

A lobe is rotated in a close fitting casing with inlet and outlet ports. The rotary motion of the lobe delivers the entrapped liquid into discharge receiver. The high pressure liquid is delivered from the receiver.

The volume displaced by the pump per second,

$$Q = 2\pi RB\frac{N}{60}\eta_v\ (\text{m}^3/\text{s})$$

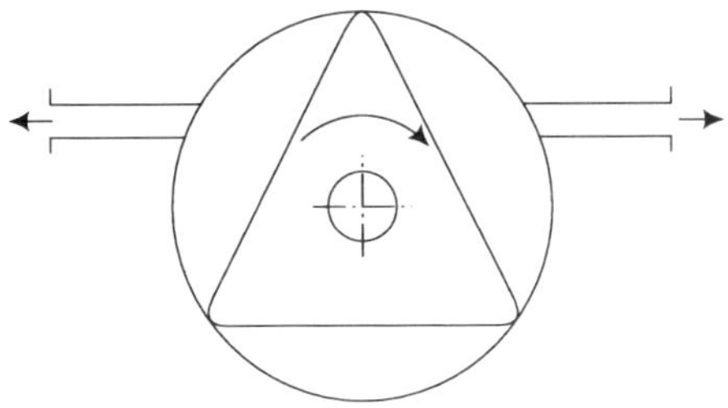

Fig. 11.9 (a) Lobe pump

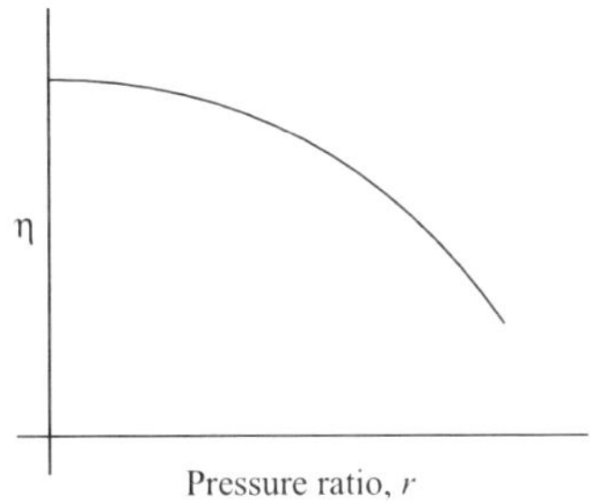

Fig. 11.9 (b) Lobe pump efficiency

where R = inner radius of casing [m]
B = width of lobe (m)
N = speed [rpm]

There are twin lobe pumps working in the same way as the gear pumps.

$$\eta_v = \text{volumetric efficiency}$$

The volumetric efficiency falls with pressure ratio.

11.5.3 Sliding Vane Pump

The pump consists of an eccentrically mounted rotor in a cylindrical casing carrying inlet and outlet ports. The rotor carries spring loaded sliding vanes. The vanes moves to-and-fro in the slots of the rotor which is rotated by a driver.

The volume of the liquid displaced by the pump per second

$$Q = [2\pi(R - e) - nt]\ 2eB\frac{N}{60} \times \eta_v\ [\text{m}^3/\text{s}]$$

where R = Inner radius of casing (m)
e = Eccentricity between rotor and casing (m)
n = Number of vanes,

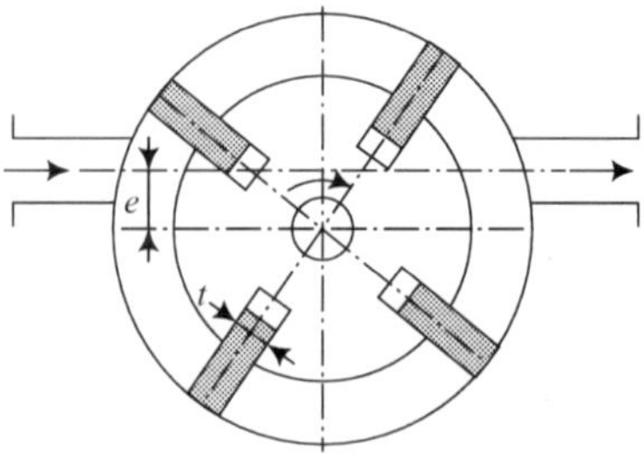

Fig. 11.10 Vane pump

t = thickness of vanes (m)
B = width of vanes (m)
N = rotor speed (rpm)
η_v = volumetric efficiency [0.7 – 0.9]

11.5.4 Screw Pump

The screw pumps are used in control and lubrication system of large drivers. These pumps have the capacity to give uniform discharge, high pressure, and operate at high speeds and without noise. However, these pumps are very expensive as compared to gear pump, lobe pumps and vane pumps for the same output. The manufacture of screws of special profiles requires complicated technology.

The pump consists of a driven worm screw meshing with two sealing worm screws rotating in a cylindrical casing with a suction and discharge chamber. As the screw rotates, the liquid is trapped in the screw flights and is moved in the axial direction and discharged.

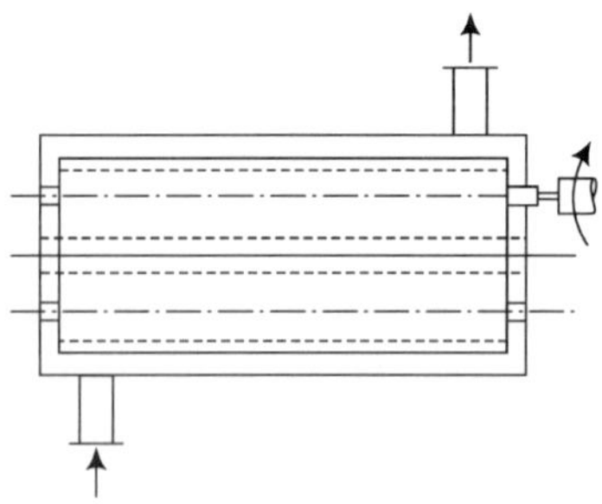

Fig. 11.11 Screw pump

The capacity of the pump is computed as:

$$Q = \frac{D^3 N}{14{,}500} \eta_v \, [l/s]$$

where N = speed of the screw [rpm]

D = diameter of worm screw [m].

η_v = volumetric efficiency [0.7 – 0.95]

11.6 JET PUMPS

Jet pumps are used to lift water from deep well or sump by the use of energy of water or compressed air.

11.6.1 Water Jet Pump

It consists of a nozzle and a diffuser mounted on a mixing chamber. Water from a centrifugal pump is supplied to the nozzle, which converts pressure energy of water into kinetic energy at the throat of the diffuser. The sump water is connected to the mixing chamber where low pressure is generated. The sump water is lifted due to pressure difference. The water mixture recovers some pressure in the diffuser and discharged outside. This pump can lift water by 6 m height.

The efficiency of jet pump.

$$\eta = \frac{Q_s}{Q_1} \times \frac{(H_s + H_d)}{(H_1 - H_d)}$$

where Q_1 = Discharge through nozzle (m^3/s)

Q_s = Discharge through suction pipe (m^3/s)

H_1 = Pressure head on delivery side (m)

H_d = Delivery head (m)

H_s = Suction head (m)

The jet pump efficiency is low of the order of 50%. The main losses occur in the mixing chamber.

The jet pump also finds applications in lubrication system of turbines where lubricating oil is pumped from oil tank to bearings of turbo-machines.

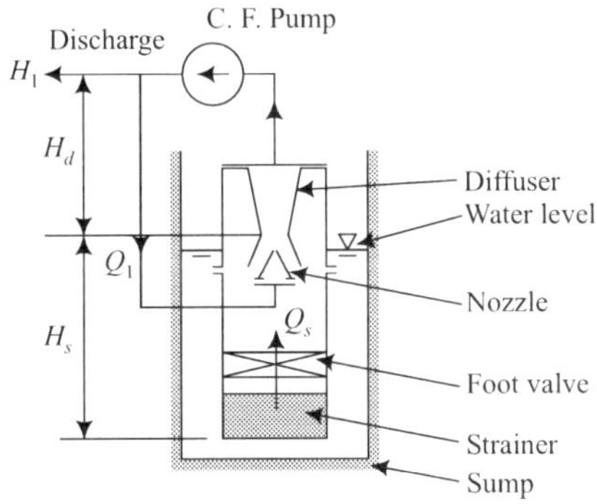

Fig. 11.12 Jet pump

11.6.2 Air Lift Pump

The air lift pumps are very simple and reliable. These are used to raise water from boreholes of small diameter and great depth. These pumps are very useful to pump corrosive liquids such as in chemical and food industries. Air lift pumps can also be used to raise liquids contaminated by sand, ash or peat.

These pumps have low efficiency of the order of 20% to 40%. Compressed air is introduced at the bottom of liquid or water column. The density of liquid air mixture is less than the density of the liquid. The liquid-air mixture rises due to density differences.

The flow rate depends upon the density difference of liquid-air mixture and the liquid.

For best results,

$$\frac{H}{h} \text{ showed lie between } \frac{1}{4} \text{ and } 1.$$

where h = submergence

H = liquid-air mixture lift

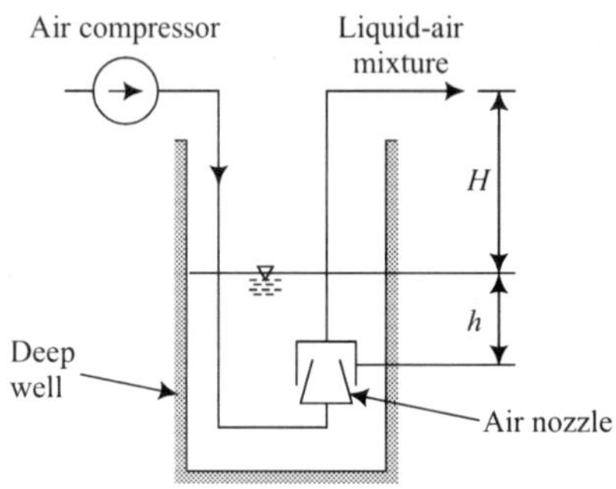

Fig. 11.13 Air lift pump

The main advantages of air lift pumps are:

1. The design is very simple due to absence of valves and moving parts.
2. Low maintenance due to absence of wear and tear.
3. Very high discharge capacity.
4. It can handle suspended solids in water safely.
5. Very suitable for drawing of water in mines where compressors are already existing.

11.7 HYDRAULIC RAM

Hydraulic ram is a pumping contrivance to raise a part of large amount of water available at some height, to a greater height. This can be employed in hilly areas where some natural source of water like a spring or a stream is available at some altitude. Work done by a large quantity of water in falling through a small height is used to raise a small part of it to a greater height. Action of water hammer makes it feasible.

No external power is, therefore, required to work this machine. Other attractions are:

1. Negligible amount of maintenance and supervision costs.
2. Continuous operation.
3. High efficiency.
4. Quiet operation.
5. Possibility of automatic adjustment of water supply.

The liftable volume diminishes asymptotically with lifting height. In case of medium lifting height, the hydraulic ram operates with an efficiency absolutely comparable to a piston pump of the same performance.

As the hydraulic ram does not need a driving unit it is ideal for lift irrigation and drinking water for hilly areas where the supply in fossil energy carriers or electricity is problematic. The hydraulic ram operates with water streams of 1 to 40 m^3/s and fall heads of 1.5 to 30 m and with lifting height upto 300 m.

Hydraulic ram is a simple and rugged device for operating irrigation schemes. This can be easily assembled from local materials.

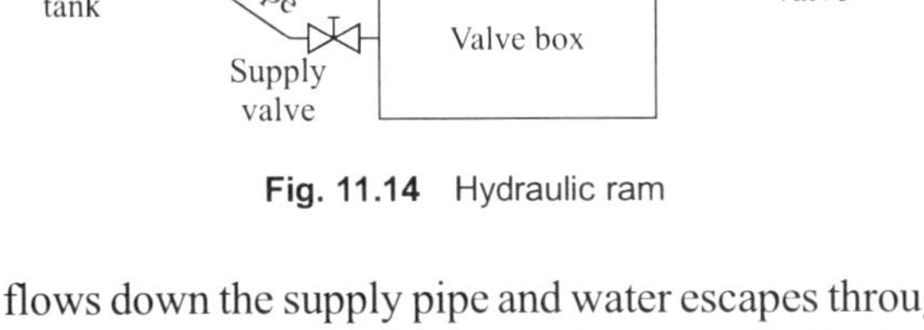

Fig. 11.14 Hydraulic ram

The water flows down the supply pipe and water escapes through waste valve. With increase of velocity of water, the dynamic pressure lifts the delivery valve which closes suddenly. The sudden stoppage of water increases pressure in the valve box and part of water enters the air vessel, thereby compressing the air. The air pressure forces the water to the delivery tank when momentum of water gets destroyed in the valve box, the delivery valve closes and waste valve opens to restart the cycle. Air vessel also helps to prevent the occurrence of water hammer in the delivery pipe and reduce fluctuations in flow rate.

The energy supplied by water from supply tank = $\rho Q_1 g H_1$

The energy supplied to discharge tank = $\rho Q_2 g H_2$

where Q_1 = water supplied to valve box (m^3/s)

Q_2 = water lifted to delivery tank (m^3/s)

The *D* Aubisson's efficiency,

$$\eta = \frac{\rho Q_2 g H_2}{\rho Q_1 g H_1} = \frac{Q_2 H_2}{Q_1 H_1}$$

The Rankine efficiency,

$$\eta_{\text{Rankine}} = \frac{Q_2}{Q_1} \cdot \frac{(H_2 - H_1)}{H_1}$$

where H_1 = Initial level of water (m)

$(H_2 - H_1)$ = Height of water lifted (m)

11.8 HYDRAULIC COUPLING

Hydraulic coupling is a combination of a pump and a turbine and is used to transmit power at constant torque. The pump unit is mounted on the driver shaft and turbine unit is mounted on the shaft of the driven machine.

The pump impeller when started imparts kinetic energy to the oil and throws outside. The oil enters the turbine runner and the turbine and driven machine starts rotating. The torque increases with the increase of driver speed and the same is imparted to the driven machine through turbine runner. Finally, the speed and torque of both the machines are same. The oil circulates in a circular path. The oil is supplied from pump impeller to turbine runner and the same is returned from turbine runner to pump impeller.

The efficiency of hydraulic coupling is defined as the ratio of power transmitted to the driven machine to the power developed by the pump impeller.

$$\eta = \frac{T_o N_o}{T_i N_i}$$

where T_o = Torque input to turbine (N-m)

T_i = Torque output of pump impeller (N-m)

N_o = Speed of driven shaft (rpm)

N_i = Speed of driver (rpm)

Neglecting friction,

$$T_o = T_i$$

$$\therefore \quad \eta = \frac{N_o}{N_i} \text{ i.e., speed ratio}$$

The slip in the speed,

$$S = \frac{N_i - N_o}{N_i} = 1 - \frac{N_o}{N_i} = 1 - \eta$$

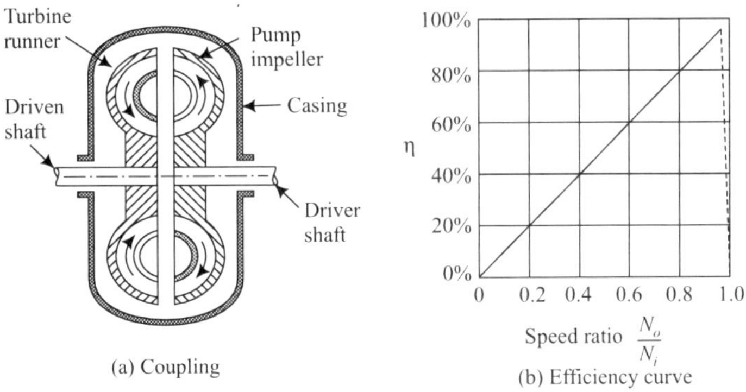

Fig. 11.15 Hydraulic coupling

11.9 TORQUE CONVERTER

Torque converter is a modified hydraulic coupling where a reacting member is added to feed variable torque to the driven machine. These are mostly used for agumentation of torque. These are widely used in automobile power transmission unit, diesel locomotives, earth moving machinery. Large magnification of torque can be obtained by using one pump impeller but two or more sets of fixed guide vanes and turbine runners.

Liquid flows from pump impeller to turbine runner through reaction chamber carrying stationary vanes which change the direction of the liquid and torque and speed transformation can be obtained.

The efficiency of torque converter,

$$\eta = \frac{\text{Power output}}{\text{Power input}} = \frac{(T_i + T_v)N_o}{T_i N_i} = \frac{N_o}{N_i}\left(1 + \frac{T_v}{T_i}\right)$$

where T_i = Torque produced by pump (N-m)

T_v = Torque of guide vanes (N-m)

N_o = speed of driven shaft (rpm)

N_i = speed of driver (rpm)

In fluid coupling, the guide vanes are absent.

$$\therefore \qquad T_v = 0$$

$$\eta = \frac{N_o}{N_i}$$

The efficiency of a torque converter is maximum when speed ratio $\frac{N_o}{N_i} = 0.5$.

Comparing the efficiency curves of torque converter and hydraulic coupling, it is clear that torque converters are more efficient at low speed ratio and hydraulic couplings are more efficient at high speed ratios. The transmission systems are so designed that the unit operates as torque converter at low speed ratios and hydraulic coupling at high speed ratios.

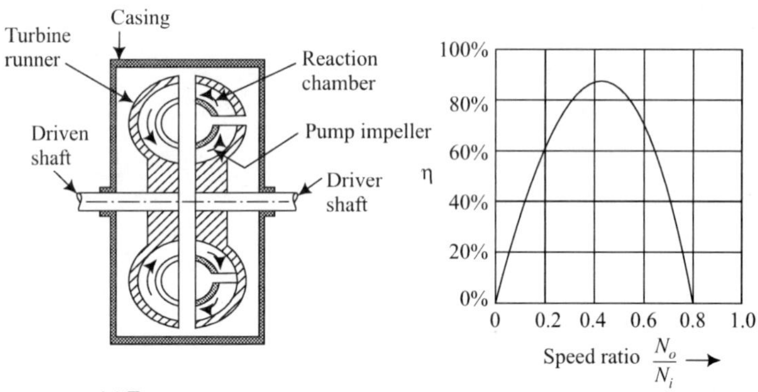

(a) Torque converter

(b) Speed curve

Fig. 11.16 Hydraulic torque converter

QUESTION BANK NO. 11

1. Explain the following aspects of an axial flow pump:
   (i) Construction
   (ii) Design constants
   (iii) Performance characteristics
   (iv) Capacity control
2. How is airfoil theory used in the design of axial flow pump? What are the limitations of this theory?
3. How is the design method based on velocity diagrams evolved from airfoil theory?

4. Explain the construction and operation of a vertical turbine pump? How is it modified for the design of a submersible pump and borewell pump?
5. Compare the construction and performance characteristics of various types of rotary displacement pumps. How is screw pump superior?
6. Write notes on:
   (a) Water jet pump
   (b) Air lift pump
7. What is hydraulic ram? How does it work? What are its advantages and disadvantages?
8. What is the difference between a hydraulic coupling and torque converter? How a torque converter can be used as hydraulic coupling?

TUTORIAL SHEET NO. 11

1. A hydraulic ram lifts water from a tank 3 m above the ram to a tank 20 m above the ram. The ram supplies 0.1 m^3/s of water and discharges 0.005 m^3/s through a 6.5 cm diameter pipe which is 80 m long. Calculate the efficiency of the ram by taking friction coefficient, $f = 0.01$. Neglect friction losses in supply pipe.

Solution The water velocity in discharge pipe,

$$V = \frac{Q}{A} = \frac{0.005}{\frac{\pi}{4}\left(\frac{6.5}{100}\right)^2} = 1.507 \text{ m/s}$$

Friction head loss

$$\Delta H_f = \frac{4fLV^2}{2gD} = \frac{4 \times 0.01 \times 80 \times (1.507)^2}{2 \times 9.81 \times \left(\frac{6.5}{100}\right)}$$

$$= 5.7 \text{ m}$$

Total delivery head = 20 + 5.7 = 25.7 m

(a) Aubisson's efficiency,

$$\eta = \frac{Q_2}{Q_1} \times \frac{H_2}{H_1} = \frac{0.005}{0.1} \times \frac{25.7}{3} = \mathbf{42.8\%} \quad \text{Ans}$$

(b) Rankine efficiency,

$$\eta_{\text{Ranking}} = \frac{Q_2}{Q_1} \cdot \frac{(H_2 - H_1)}{H_1} = \frac{0.005}{0.1} \times \frac{(25.7 - 3)}{3}$$

$$= \mathbf{37.8\%} \quad \text{Ans}$$

PART IV

SYSTEMS AND PLANTS

CHAPTER 12

Hydraulic Systems

12.1 INTRODUCTION

The lifting machines like hydraulic lifts and cranes and jacks and presses, etc. are operated by high pressure oil or water by a pump working continuously. The energy requirement of these devices may not be uniform. Large amount of energy is needed when the load is hoisted and practically no energy is needed when the load is lowered. The non-uniform energy requirements can be met by the following methods:

1. A large capacity pump with intermittent operation.
2. A small capacity pump with hydraulic accumulator and pump working continuously.

Energy is supplied from pump plus energy of the accumulator during load lifting. However, when no energy is needed during lowering of load, pump energy is stored in the accumulator.

The energy requirement of a lifting device may be enormous and cannot be met by the pump. An intensifier may be used to multiply the pump pressure to supply high pressure to the lifting device. The layout of a system for a hydraulic lifting device may consist of a pump, an accumulator and intensifier as shown in Fig. 12.1.

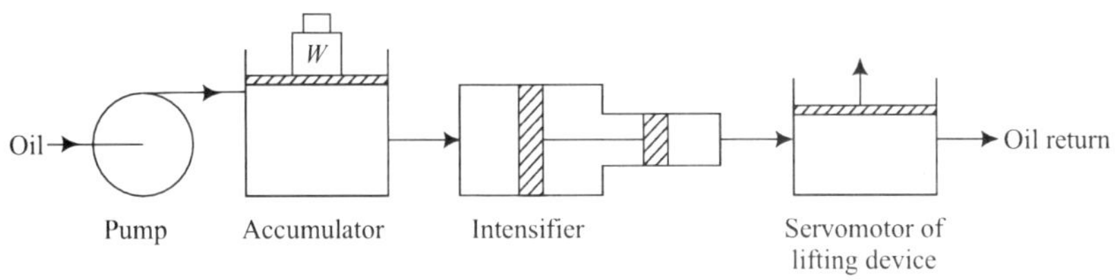

Fig. 12.1 Hydraulic system for lifting device

1. *Pump:* The working fluid can be high pressure water or oil. Oil is more widely used which is supplied at high pressure by a rotary positive displacement pump. Gear pump or screw pump are widely used.

2. *Accumulator:* A simple or differential accumulator is used to store the pressure energy of the pump.
3. *Intensifier:* Hydraulic, hydro-pneumatic or steam intensifiers are used to multiply the oil pressure.
4. *Servomotor:* The lifting devices are operated by a hydraulic servomotor which is a piston-cylinder unit and actuates or opens a device by piston motion controlled by slide or pilot valve.

12.2 HYDRAULIC ACCUMULATOR

Hydraulic accumulator is a device to store energy is the form of liquid pressure and to release the pressure when required.

The capacity of the accumulator is the maximum amount of energy, it can store.

The capacity of the accumulator = ApL [N-m]

where A = Area of ram [m^2]

L = Maximum stroke of ram [m]

p = Liquid pressure [N/m^2]

The force or load lifted $W + W_f = AP$ [N]

where W = Dead weight of ram plus external weight [N]

W_f = Force of friction between ram and cylinder [N]

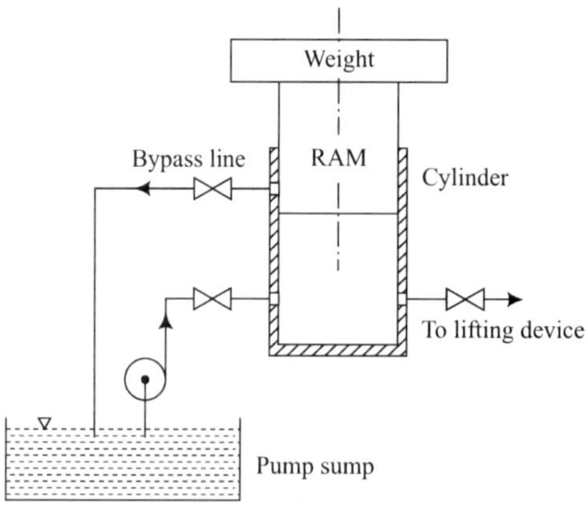

Fig. 12.2 Hydraulic accumulator

12.3 DIFFERENTIAL ACCUMULATOR

Differential accumulator is used when the capacity handled by the accumulator is small but pressure requirement is high.

Liquid under pressure from the pump enters the sliding cylinder cum weight through a hole in the lower fixed ram. The cylinder is lifted till it is balanced by its weight. During downward movement of the loaded piston, the liquid accumulated is pushed to the lifting machine.

The area of lower fixed piston is more than area of upper fixed piston,

$$A_2 > A_1$$

Capacity of the accumulator $= (A_2 - A_1)\,Lp$

where L = stroke of piston

$$\text{Capacity} = W \times L$$

where W = weight of sliding piston.

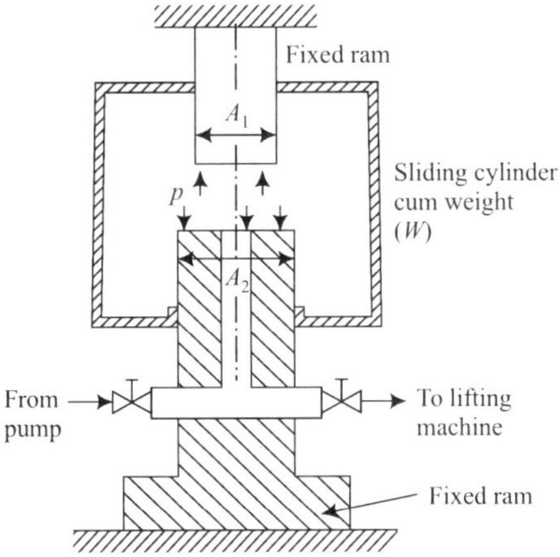

Fig. 12.3 Differential accumulator

12.4 HYDRAULIC INTENSIFIER

Hydraulic intensifier is used to multiply the pressure of the pump to supply high pressure to the lifting machine. It is differential piston-cylinder device. The low pressure liquid or air or steam is supplied to the larger cylinder of piston area A_1. This pushes the piston of area A_2 in the small cylinder where working fluid is supplied from the reservoir.

The pressure p_2 is more than pressure p_1.

The forces acting on the differential piston are equal due to equilibrium

$$F_1 = F_2$$

$$\therefore \quad A_1 p_1 = A_2 p_2$$

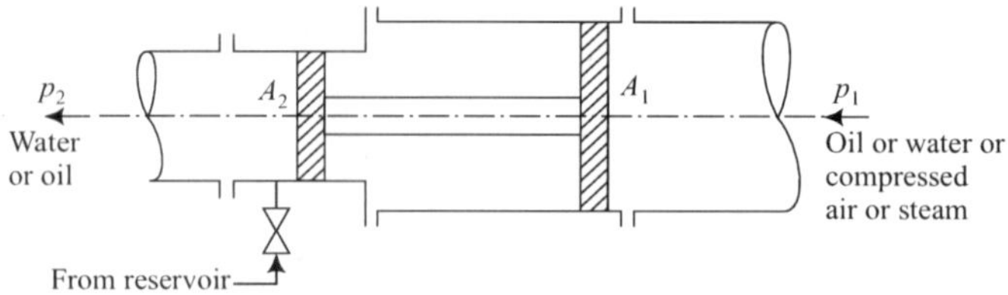

Fig. 12.4 Hydraulic intensifier

$$\therefore \quad p_2 = p_1\left(\frac{A_1}{A_2}\right) \quad \text{(Neglecting friction resistance)}$$

There are single acting intensifiers and also double acting intensifiers for continuous operations. Pressures of the order of 1500 bar have been achieved by intensifiers.

In hydraulic intensifiers water or oil is used. In hydro-pneumatic intensifiers, the fluid supplied is compressed air. Steam is also used in some intensifiers.

12.5 HYDRAULIC JACK OR PRESS

Hydraulic jack is a portable equipment to lift heavy loads over short distance.

Hydraulic press is a permanent set-up used for pressing cotton bales in ginning mills, plasting moulding machines and metal pressing work, for pushing and shearing steel plates. An effort applied to the pump plunger (d) through lever arrangement produces a pressure p in the hydraulic liquid. This pressure acts on the ram (D) which supports the load (W). The load

$$W = \frac{\pi}{4}D^2 p$$

The rate of lifting of load will be equal to the rate of oil supplied by the pump. For lowering the load, pressure relief valve is opened to drain the oil into reservoir.

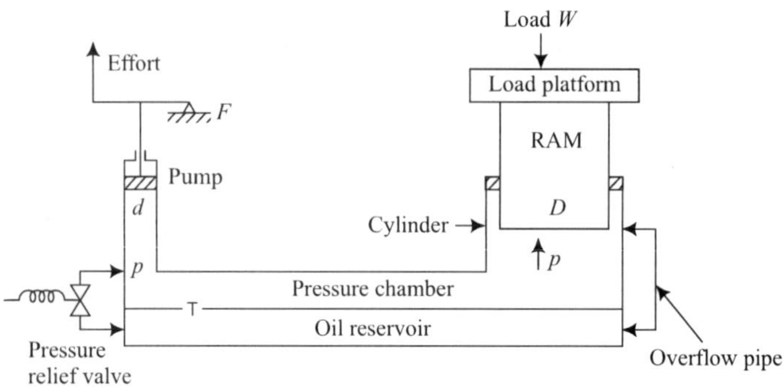

Fig. 12.5 Hydraulic press

12.6 HYDRAULIC LIFT

Hydraulic lift is used to carry goods and passengers from one floor to another in multi-storeyed buildings. In general, hydraulic lifts have been superceded by electric lifts and hydraulic lifts are used as standby units wherever there is danger of fire and explosion.

Fluid under pressure is supplied to a cylinder under a ram carrying a cage or platform at its top. In *direct acting* lifts, the stroke of ram is equal to one floor height. Modern lifts are generally of *suspended* type.

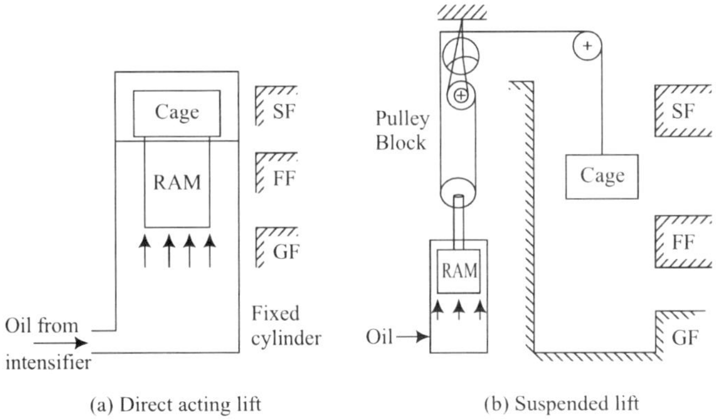

Fig. 12.6 Hydraulic lift

QUESTION BANK NO. 12

1. With the help of a flow diagram, list the various components of hydraulic system for lifting device.
2. What is a hydraulic accumulator? Differentiate between an ordinary and differential accumulator.
3. Explain the working of the following devices.
   (a) Hydraulic Intensifier
   (b) Hydraulic Press
   (c) Hydraulic Lift

TUTORIAL SHEET NO. 12

1. A certain machinery is operated with a hydraulic accumulator through a pipe 180 m long and 100 mm diameter. The accumulator has ram 250 mm diameter and 3 m stroke and is loaded with 345 kN weight. It is fed with water from a three-throw pump at 45 rpm. The plunger of the pump has a diameter of 45 mm and stroke of 350 mm; the slip being 3 per cent. If the

machinery absorbs 36.8 kW power, calculate the longest period for which it can be operated continuously. Take friction fact or, $f = 0.032$.

Solution Intensity of pressure in the accumulator,

$$p = \frac{W}{A} = \frac{345 \times 10^3}{\frac{\pi}{4}(0.25)^2} = 7028 \times 10^3 \text{ N/m}^2$$

Volume of accumulator $= AL$

$$= \frac{\pi}{4}(0.2)^2 \times 3 = 0.147 \text{ m}^3$$

Friction losses in the pipe,

$$\Delta H_f = \frac{4fLV^2}{2gD} = \frac{4 \times 0.032 \times 180 \times V^2}{2 \times 9.81 \times 0.10}$$

$$= 11.74\ V^2 \text{ m}$$

Available energy at the end of pipe,

$$\rho QgH = \left[\frac{7028 \times 10^3}{\rho g} - 11.74V^2\right]\rho Qg$$

$$= (716 - 11.74\ V^2)\rho \times 9.81 \times \frac{\pi}{4}(0.1)^2 V \text{ N/m/s}$$

Energy required for the machinery = 36.8 kW

Equating the two

$$(716 - 11.74\ V^2) \times 0.077\ V = 36.8$$

$$\text{V} = 0.67 \text{ m/s}$$

Flow at the end of pipe,

$$Q = \frac{\pi}{4}(0.1)^2 \times 0.067 = 0.00526 \text{ m}^3\text{/s}$$

Flow from pump,

$$Q_p = \frac{A_p L_p N_p}{60} \times 3 \times \eta_v$$

$$= \frac{\pi}{4}(0.045)^2 \times 0.35 \times \frac{45}{60} \times 3 \times 0.97$$

$$= 0.001215 \text{ m}^3\text{/s}$$

Flow required from accumulator,

$$Q_{acc\cdot} = Q - Q_p = 0.00526 - 0.001215$$
$$= 0.004045 \text{ m}^3/\text{s}$$

$$\text{Time of supply by accumulator} = \frac{0.147}{0.004045} = \mathbf{36.6 \text{ s}} \quad \text{Ans}$$

2. An intensifier has a ram diameter of 150 mm and a sliding cylinder diameter of 570 mm. Calculate the pressure of water on the low pressure side of the intensifier, if the pressure of water on the high pressure side is 20 MN/m^2.

Solution For an intensifier

$$p_1 A_1 = p_2 A_2$$

$$\therefore \qquad p_2 = p_1\left(\frac{A_1}{A_2}\right) = p_1\left(\frac{\frac{\pi}{4}D_1^2}{\frac{\pi}{4}D_2^2}\right)$$

$$= 20 \times \left(\frac{750}{150}\right)^2 = \mathbf{500 \text{ MN/m}^2} \quad \text{Ans}$$

3. A hydraulic accumulator has 200 mm ram diameter and lift of 500 mm. Calculate the capacity of the accumulator when it is supplied with water at a pressure of 35 bar.

Solution Capacity of accumulator = ApL

$$= \frac{\pi}{4}D^2 pL$$

$$= \frac{\pi}{4}(0.2)^2 \times 0.5 \times 35 \times 10^5$$

$$= \mathbf{0.35 \times 10^2 \text{ kN-m}} \quad \text{Ans}$$

4. A hydraulic lift has a ram diameter of 200 mm and is supplied with water at a pressure of 20 bar. Find the total load it can lift if the velocity of lift is 1 m/s. Assuming efficiency of 60%, calculate the power required.

Solution The load,

$$W = \frac{\pi}{4}d^2 p = \frac{\pi}{4}(0.2)^2 \times 20 \times 10^5$$
$$= 632832 \text{ N}$$

The power required,

$$P = \frac{W \times t}{\eta} = \frac{632832 \times 1}{0.6} = \mathbf{105.5 \text{ kW}} \quad \text{Ans}$$

CHAPTER 13

Hydraulic Power Plants

13.1 INTRODUCTION

The utilization of hydro energy to operate agricultural and industrial devices is one of the oldest and widespread techniques of human achievement. Hydro power, available free of cost in the form of flowing water, is an important, pollution free and renewable source of energy. In a hydroelectric power plant, water streams in hills/ mountains are collected behind a dam. The potential energy of stored water is passed to a hydro turbine where potential energy of water is used to rotate the turbine runner. The mechanical energy of rotation of turbine is used to run an electrical generator to produce electricity. The electricity generated is transmitted over distances and distributed to consumers of electrical energy.

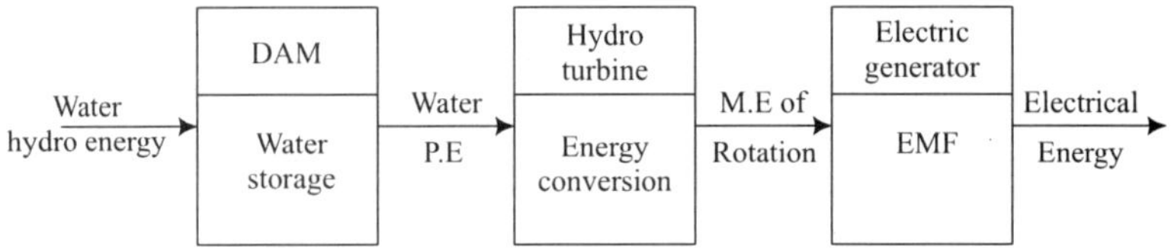

Fig. 13.1 Hydro power plant

13.2 HYDRO POWER POTENTIAL

The hydro power potential of the world is huge to the tune of 36×10^{12} kWh of annual electricity generation which is technically feasible. Approximately 11×10^{12} kWh of electricity can be economically generated every year. Presently 20% of the total world electricity is generated by hydroelectric power stations. In some countries, the share of hydroelectric power is very high as seen in Table 13.1.

The Central Electricity Authority (CEA) has assessed India's hydro potential to be about 1,48,700 MW of the installed capacity. The hydro electric capacity currently under operation is about 26,000 MW and 16,083 MW is under various stages of development. India is endowed with economically viable hydro potential

Table 13.1 Share of hydro power in various countries

Sl. No.	Country	Share of Hydropower	Sl. No.	Country	Share of Hydropower
1.	Norway	99%	6.	India	25%
2.	Brazil	94%	7.	USA	13%
3.	Canada	66%	8.	Former USSR	14%
4.	Austria	69%	9.	Germany	5%
5.	Switzerland	62%			

of 600 TWH (trillion Watt hours). The key advantage of hydroelectric power is the ability to store energy and the flexibility of its use during peak load periods.

13.3 CLASSIFICATION OF HYDRO POWER PLANTS

Hydro power plants are capital intensive projects with long construction period. These are mostly designed as multi-purpose integrated plants. In addition to generation of electricity, these projects are designed for irrigation water and flood control.

Broadly, these plants are classified as run-of-the-river plant, plants with storage and pumped hydro plants. Depending upon the power output, hydro power plants are classified as small (upto 10 MW), medium (upto 100 MW) and large power plants (above 100 MW). There are also mini, micro and small power plants with capacities in the range of 100 kW to 1 MW. Besides, hydroelectric power plants are classified as low-head (upto 10 m), medium-head (upto 100 m) and high-head (above 100 m) plants.

13.3.1 Run-of-River Plants

These power plants are installed on large and small rivers and irrigation canals. These are low head and high water flow power plants. The water head is created with the help of a dam or a weir. A water supply channel runs parallel to the river with a low slope on which powerhouse is built.

The power generated depends on the quantity of flow water and works on daily basis on the amount of flow available. Sometimes, a small reservoir or pond is built, which can store a few hours supply of water to the plant, when the river flow exceeds the amount required by the plant. The stored water is used in generating power during the hours when the demand is in excess of the flow of water in the river at the moment.

Run-of-river plants are generally low-head small power plants using Propeller or Kaplan turbines or bulb or tubular turbines with water siphoning over the weir.

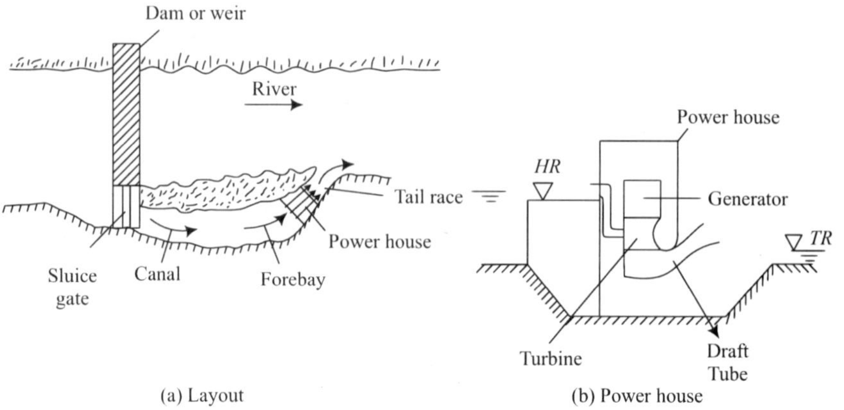

Fig. 13.2 Run-of-river power plant

13.3.2 Plants with Storage Reservoir

High head power plants work in high mountains where water is stored in natural lakes during rainy season or when the snow melts. Long penstocks are used to connect the lakes to powerhouses where Pelton turbines are installed. These are mostly medium capacity plants.

The most common power plants in India are plants with storage reservoir. These are built on large rivers by building a dam. During rainy season, water is stored in reservoirs behind the dams so that it can be utilized during other seasons to supplement the flow of river whenever the flow in the river falls below a specified minimum based on study of flow duration curve. Power can be generated directly from the reservoir. Sometimes canals are constructed to convey water from the reservoir for irrigation purposes.

Storage power plants are designed for medium head and Francis turbines. The size and number of turbines are decided by optimization of hydrothermal mix by interconnecting hydro power with steam or nuclear power plants.

The storage reservoir behind the dam can store water for a very long period of time. These plants cover both peak and base electrical load.

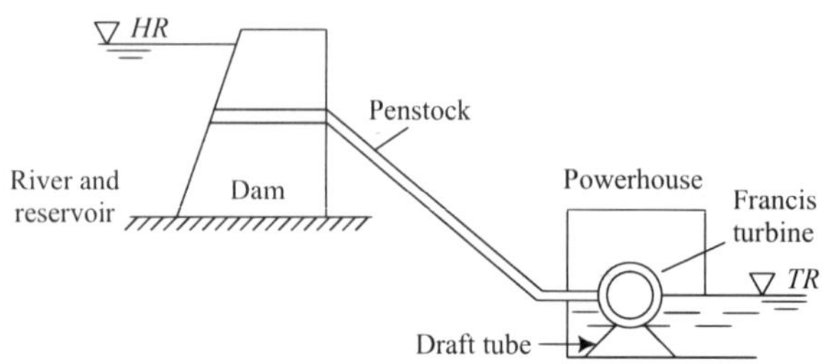

Fig. 13.3 Storage reservoir power plant

The largest storage reservoir power plants in the world are listed in Table 13.2. along with their installed capacities.

Table 13.2 Storage reservoir hydro power plants

Sl. No.	*Name of station*	*River and country*	*Installed capacity (MW)*
1.	Itapu	Parana River Brazil/Paraguay	12,600
2.	Grand Coulee	Columbia River, USA	10,800
3.	Guri	Canoni River, Venezuela	10,300
4.	Sayano-Shushearskaya GES	Yenisseri River, Russia	6,400
5.	Khasnoyarskaya GES	Yenisseri River, Russia	6,000

The largest power plants in India are Bhakra Dam at Satluj River and Tehri Dam at Ganga River.

13.3.3 Pumped Storage Plants

Pumped hydro storage plants are mainly used to meet peak load demand. These can also be used as standby power plants to quickly start in case of breakdown of a thermal or nuclear plant working as a base load power plant. The start-up time of pumped storage plant is approximately 1 minute. These are mostly high head plants. The plant consists of a number of turbine-generator sets and pump-motor sets.

Water after working in turbine is stored in tailrace reservoir. During *low load*, say night time, the water is pumped back from the tail to head reservoir drawing excess power from the grid or from nearly base load thermal or nuclear power plant. During *peak load*, the water is used to work on turbines to produce electricity. The steam power plants are basically base-load plants working at full plant capacity.

Whenever the load demand is less than full plant capacity, the surplus power is used to run the pump installed at the tail race of hydroelectric plant. Storing of

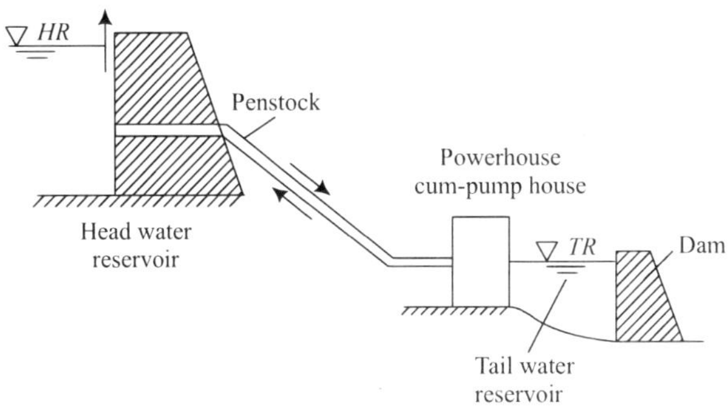

Fig. 13.4 Pumped storage plant

potential energy of water is the most cost-effective and environment-friendly method of energy storage. About 75 per cent of power used in pumping is recovered. A motor-generator set and a reversible turbine-pump unit can be profitably used as the same machine can be operated as a motor or generator and similarly as a pump or turbine (Deriaz turbine).

The specifications of world's largest pumped hydro plants are given in Table 13.3.

Table 13.3 Specifications of pumped storage plants

Sl. No.	*Power station, country*	*Head (m)*	*No. of sets*	*Installed capacity (MW)*	
				Turbines	*Pumps*
1.	Bath country, USA	360	6	2100	2280
2.	Luding town, USA	110	6	1755	1910
3.	Dinorwig, U.K.	535	6	1620	1800
4.	Raccoon, USA	320	4	1530	1600
5.	Shinu Takase-gawa, Japan	265	4	1280	1320
6.	Grand Maison, France	955	8	1220	1260
7.	Oku-Yashino, Japan	515	6	1210	1280
8.	Zagork, Russia	1120	6	1200	1250
9.	Piastra, Italy	610/1070	9	1280	1430
10.	Weir Hornberg, Germany	660	4	960	1040

The energy flow diagram of a pumped hydro plant is shown in Fig. 13.5. From electric energy consumed for pumping 75% is regained in hydro-generator sets.

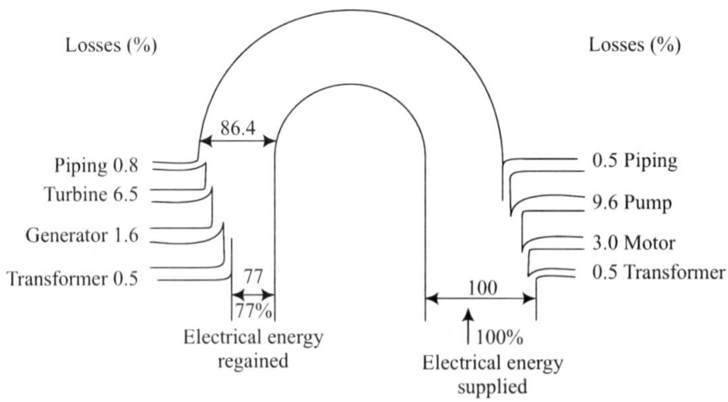

Fig. 13.5 Energy flow diagram for pumped storage plant

13.3.4 Small, Mini and Micro Hydro Power Plants

Small hydro power plants have no impact on environment and are safe for fish breeding and drinking water supply. The natural water sources in hilly areas can be utilized for power generation with low-head standardized hydro sets. The range of applications of hydro turbines for small power plants is given in Table 13.4. Small and micro hydro systems are very important for developing countries which possess large hydro resources.

Table 13.4 Operating range of hydro turbines for small power plants

Sl. No.	*Type of Turbine*	*Power Capacity*	*Head (m)*
1.	Pelton	upto 5 MW	upto 450
2.	Francis	750 – 6000 kW	25 – 160 m
3.	Propeller	upto 220 kW	Below 25 m

The following countries have following operational small hydro plants.

China	90,000
USA	10,500
Japan	5,300
Switzerland	2600
Austria	1300

The potential energy source in India for small plants is around 20,000 MW. The mini plants are operating with 5 m to 20 m head with installed capacity of 1 MW to 5 MW. The micro plants are operating with head less than 5 m and with installed capacity between 0.1 MW and 1 MW.

13.4 LOCATION OF HYDRO POWER PLANTS

The following important selections have to be made for the design of hydraulic power plants.

1. Location

The site selection must ensure economical availability of all resources, water flow, head, accessibility, reservoir area, dam foundation, etc.

2. Installed capacity

The installed capacity is decided by hydrology of the site, i.e., hydrographs, flow duration curves, range of available head, etc.

3. Number and size of units

The number and size of hydro sets will be decided by load duration curves and interconnected operation with other power plants, i.e., steam and nuclear power plants. The type of turbines will be decided on the basis of flow, head, specific speed, size and availability of hydro sets from suppliers.

The following factors are considered for the location of a hydro power plant.

1. Availability of water

The design and capacity of the hydro power plant depends upon the amount of water available at the site. The data on run-off, annual precipitation in the catchment area, maximum and minimum quantity of water available in a year is required to decide the capacity of the plant, integration with thermal and nuclear plants and waterworks for flood control and irrigation water.

2. Water storage capacity

The storage capacity has to be decided to ensure continuous generation of power as there can be wide variation in the quantity of water flow in the river and rainfall round the year. The storage capacity is estimated with the help of hydrographs and flow duration curves.

3. Available water head

The power generation is a function of available water flow rate as well as available head. Availability of large head reduces the quantity of water requirement for a given output.

4. Accessibility of site

Large quantity of men and materials have to approach the site during long construction and operation periods. The site should be easily accessible by rail and road.

5. Distance from load centre

The power generated has to be transmitted to load centres. Therefore, the site should be close to load centre to reduce the cost of transmission lines as well as transmission losses.

6. Type of land

Many types of structures have to be built for the hydro power plant such as dam, conduits, intakes, surge tank, powerhouse, head and tail reservoirs. The land should be cheap and rocky. It should be strong enough to withstand the stress in the structure and the thrust of water.

7. Project cost

The cost of the project and the completion period should be minimum. Detailed investigations are carried out which include:

1. *Hydrological investigations* for water availability, storage capacity, available head, distance from load centre and approach to site by road and rail.
2. *Topographical investigation* of the catchment area and site for building of dam, conduit, powerhouse with the help of level contours by ground surveying or aerial surveying.

3. *Geological investigation* for accurate information of ground characteristics. The selection of site depends on the following factors:
   (i) Tight basin of ample size
   (ii) Narrow outlet requiring dam of low volume.
   (iii) Strong foundation to support the dam structure.
   (iv) Safe and ample spillway for disposal of surplus water.
   (v) Valuable minerals and agricultural lands should not be submerged.
   (vi) Local availability of materials for construction of dam.
4. *Water pollution effects:* The project should have minimum impact on the surroundings and enhance the local environment. The sedimentation effect of the reservoir should be minimum as it can lead to reduction of reservoir capacity and erosion of turbine blades.

13.5 HYDROLOGY

Hydrology is a science which deals with depletion and replenishment of water resources over and under the earth's surface. It is used to design irrigation and flood control schemes, power projects, water supply system, navigation work, etc.

The precipitation in the form of rain, hail, snow or sleet falls on the hills and mountains and converges to form streams and rivers. This can be used for power generation. The whole of precipitation or rainfall does not convert into river flow called *run-off*. There are following losses of water:

1. *Re-evaporation* of part of precipitation into atmosphere.
2. *Transpiration* of water through roots of plants and through leaves into atmosphere.
3. Infiltration of water by seepage into the ground forming water table or ground storage.

The difference of rainfall in the catchment area of a river and the losses is called run-off and is available for power generation.

The hydraulic power.

$$P = \rho Q g H \eta \text{ [W]}$$

where ρ = Water density [1000 kg/m^3]

g = Acceleration due to gravity [9.81 m/s^2]

Q = Flow of water [m^3/s]

H = Head of water due to height of fall of water [m]

η = Overall efficiency of power plant (0.85 – 0.90)

The electrical energy produced

$$W = \frac{\rho Q g H \eta t}{1000} \text{ [kWh]}$$

where t = operating time of plant [8760 h/year]

13.5.1 Flow Duration Curve

The stream flow at a site is variable as it depends on the amount of rainfall which is seasonal. The variation of the river flow with time is shown graphically as hydrographs.

A flow duration curve is a very useful tool for estimation of hydraulic power. It is drawn with the help of hydrographs. The flow duration curve is drawn between quantity of river flow against percentage of time duration for which a certain quantity of flow is available. The area under the curve gives the total quantity of run-off in a given period.

For a given head

$$P \propto Q$$

The flow duration curve can be easily converted into power duration curve by changing the units of y-axis from m^3/s to kW.

The minimum flow rate Q_m is available all the time (for 100% time) and point P_B gives *primary power* or firm power.

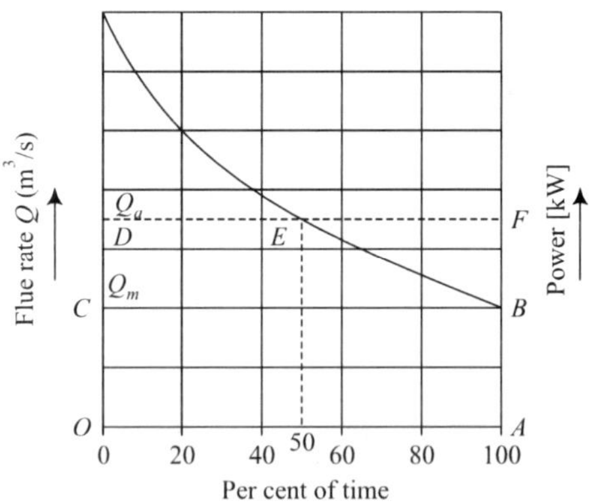

Fig. 13.6 Flow duration curve

The horizontal line *DEF* is drawn for power (P_F) required all the time. The additional power ($P_F - P_B$) is called *secondary power*. This power can be possible if additional water flow Q_a is made available. This is possible only by providing adequate storage equal to area *BEF*.

The uses of flow duration curve are as follows:

1. Preliminary studies of a hydro project
2. Evaluation of low level flows
3. Estimation of primary power (without storage) and secondary power (with storage).
4. Preparation of power duration curve.

13.5.2 Mass Curve

A mass curve is a graphical representation of accumulated flow against time. The slope of the line at any time shows the rate of flow at the time. The time is shown in years and accumulated flow in hectare-metre. The data is prepared from the records of monthly flows of a river or stream.

The mass curves are used to estimate the water storage requirements. The following characteristics should be noted.

1. If the mass curve is horizontal, the flow rate is zero.
2. Concave depression indicates dry periods.
3. A steep rise of curve indicates high flow rate.

The mass curve for a river is shown in Fig. 13.7.

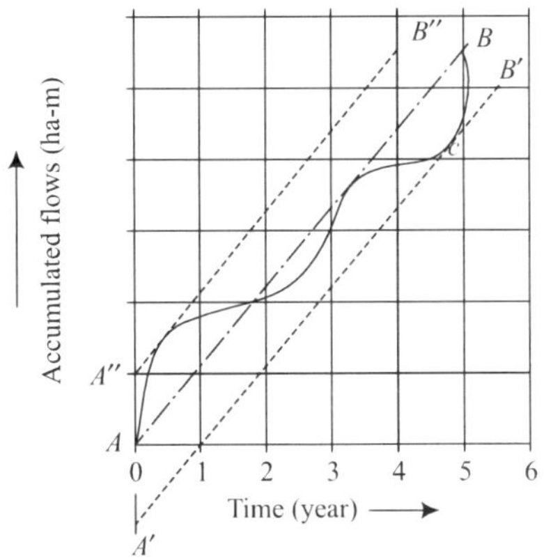

Fig. 13.7 Mass curve

Line *AB* is drawn by joining the end points of mass curve. It indicates the average discharge over the total period.

Line *A′B′* is drawn parallel to line *AB* and tangent to lowest point of mass curve *C*.

Line *A″B″* is drawn parallel to line *AB* and tangent to highest point *D* of mass curve.

The vertical intercept $A'A''$ gives the storage volume required to ensure continuous release of water for average discharge rate over the entire period.

Storage is impounding of excess run-off during the seasons of surplus flow for use in dry season. A dam is constructed across the river at a suitable site and storage is built on the upstream side of dam.

13.6 COMBINED OPERATION OF DIFFERENT POWER PLANTS

A hydraulic power plant can be used as an exclusive source of power. However, the output of the plant is proportional to water flow rate which is seasonal. During rainy season there will be plenty of water. If the plant is designed for minimum flow rate, there will be great wastage of water over the dam for most part of the year. During winter season, there will be low water flow rate and hydro power plant may not be able to meet the maximum demand. A large amount of water has to be stored to meet the variable load requirements. This is a big source of ecological imbalance claiming large amount of fertile land, damage to forestry and evacuation and rehabilitation of large population.

In order to meet the fluctuating load requirement efficiently, economically, reliably and continuously the hydroelectric power plants are used in conjunction with thermal power plants and nuclear power plants as an interconnected system. Load sharing by hydro plant is maximum when available water flow is maximum (during rainy season). As long as there is plenty of water available stored in the reservoir, the hydro plants can carry base load, with thermal plants taking the peak loads. When water flow rate is low (during winter season), the steam plants can operate as base load and hydroplants meet the peak load.

There are many advantages of interconnected mode of operation.

1. The reliability of continuous supply of electricity to the consumer is much greater in interconnected than in isolated system. The reliability of hydro plants is dependent on water flow rate and storage availability.
2. The installed capacity requirement is low because there is certain diversity in time between peak demands of two systems. Therefore, peak load of the interconnected system is less than the sum of individual peaks.
3. The power supply is ensured in the event of failure of one power station as consumers are fed from common grid of interconnected power plants.
4. The spinning spare requirement of interconnected system is reduced.
5. The overall cost of energy per unit is less from interconnected system.
6. Integrated system requires less capital investment and less operation and maintenance cost.
7. Integrated system facilitates better use of transmission lines at higher voltage by teeing the generating station in a group.

The load division between a hydro power plant and steam power plant is shown in Fig. 13.8.

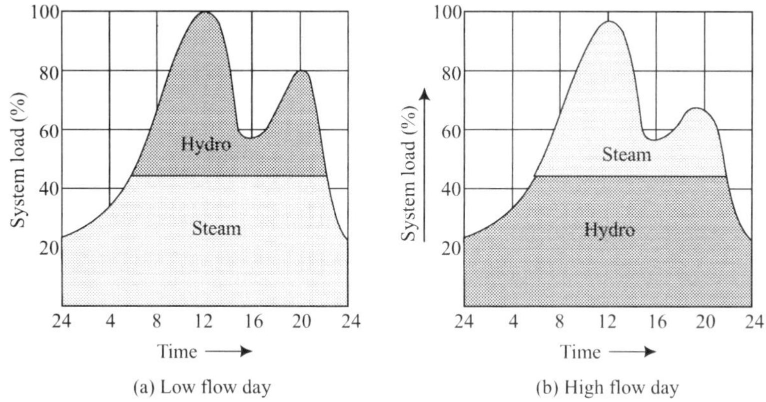

Fig. 13.8 Hydro thermal mix

QUESTION BANK NO. 13

1. Sketch the hydro power potential for the world and India.
2. Give a detailed classification of hydro power plants. Which type of hydro power plants have maximum scope in India?
3. What are crucial factors considered for the location of a hydroelectric power plant?
4. Define hydrology, flow duration curve, mass curve. How are these tools used for the design of hydro power plant?
5. What are the advantages of interconnected mode of operation of hydroelectric and thermoelectric power plants?
6. What concepts are used in the division of load between a hydro power plant and steam power plant?

TUTORIAL SHEET NO. 13

1. The following data is available for the design of a hydroelectric power plant
    1. Head available = 27 m
    2. Catchment area = 430 sq.km
    3. Rainfall = 150 cm/year
    4. Utilization of total rainfall = 65%
    5. Penstock efficiency = 95%
    6. Turbine efficiency = 80%
    7. Generator efficiency = 86%
    8. Load factor = 0.45

    Calculate the installed capacity of power plant. Also select suitable turbine.

Solution Total quantity of annual water available

$$= (430 \times 10^6) \times 1.5 \times 0.65 = 419.25 \times 10^6 \text{ m}^3$$

Available water flow rate,

$$Q = \frac{419.25 \times 10^6}{365 \times 24 \times 60 \times 60} = 13.29 \text{ m}^3/\text{s}$$

Power developed

$$P = \rho Q g H \eta_T \eta_t \eta_G$$

$$= 1000 \times 13.29 \times 9.81 \times 27 \times 0.95 \times 0.8 \times 0.86 \times 10^{-3}$$

$$= \textbf{2300 kW} \quad \text{Ans}$$

$$\text{Load factor} = \frac{\text{Average load}}{\text{Peak load}} = 0.45$$

$$\therefore \quad \text{Peak load} = \frac{2300}{0.45} = 5111 \text{ kW}$$

Use two turbines of equal capacity.

$$\text{Turbines capacity} = \frac{5111}{2} = 2555.5 \text{ kW}$$

As head is low, use two Kaplan turbines of 3000 kW each.

2. A power system has a maximum load of 80 MW at 35% load factor. The load can be supplied by any of the following schemes.
   (a) A steam power plant to supply the whole load.
   (b) A hydel plant and an interconnected steam plant. The steam plant supplies 120×10^6 kWh/year with a maximum load of 50 MW.

   The cost data is tabulated below.

Cost	*Steam Plant*	*Hydro plant*
1. Capital cost	Rs 600/kW	Rs 1400/kW
2. Operating cost	4.8 paise/kWh	1.2 paise/kWh
3. Transmission	Negligible	0.3 paise/kWh
4. Interest and depreciation	12%	10%

Calculate the energy cost for the mode of operation.

Solution Annual energy requirement

$$= \text{Maximum load} \times \text{Load Factor} \times 8760$$

$$= 80 \times 10^3 \times 0.35 \times 8760 = 245 \times 10^6 \text{ kWh}$$

(a) *Isolated Steam Plant*

Fixed cost (interest and depreciation)

$$= \frac{12}{100} \times 80 \times 10^3 \times 600 = \text{Rs. } 576 \times 10^4/\text{year}$$

$$\text{Operating cost} = \frac{4.8}{100} \times 245 \times 10^6 = \text{Rs. } 1152 \times 10^4/\text{year}$$

$$\text{Total annual cost} = \text{Rs. } (576 + 1152) \times 10^4 = \text{Rs. } 17.28 \times 10^6$$

$$\text{Energy cost} = \frac{17.28 \times 10^6}{245 \times 10^6} \times 100 = \mathbf{7.05\ paise/kWh} \quad \text{Ans}$$

(b) *Interconnected Power System*

(i) *Hydel Plant*

$$\text{Annual fixed cost} = \frac{10}{100}(80 - 50) \times 10^3 \times 1400 = \text{Rs. } 4.2 \times 10^6$$

Energy supplied by hydel plant = $(245 - 120) \times 10^6 = 125 \times 10^6$ kWh

Operating cost including transmissions

$$= 125 \times 10^6 \times \frac{(1.2 + 0.3)}{100} = \text{Rs. } 1.875 \times 10^6/\text{year}$$

(ii) Steam power plant

$$\text{Load share} = \frac{120 \times 10^6}{8760} = 13.7 \times 10^6 \text{ kW}$$

$$\text{Minimum plant capacity} = \frac{13.7 \times 10^6}{0.35}$$

$$= 39.15 \times 10^3 \text{ kW}$$

$$\text{Annual fixed cost} = \frac{12}{100} \times 39.15 \times 10^3 = \text{Rs } 2.82 \times 10^6$$

$$\text{Annual operating cost} = 120 \times 10^6 \times \frac{4.8}{100} = \text{Rs. } 5.76 \times 10^6$$

Total cost of both the plants

$$= \text{Rs. } 4.2 \times 10^6 + 1.876 \times 10^6 + 2.82 \times 10^6 + 5.76 \times 10^6$$

$$= \text{Rs. } 14.655 \times 10^6/\text{year}$$

$$\text{The energy cost} = \frac{14.655 \times 10^6}{245 \times 10^6} \times 100$$

$$= \mathbf{5.98\ paise/kWh}$$

The interconnected operation of steam and hydel station is more economical.

PART V

CASE STUDIES

CASE I

Floating Type Micro Hydro Power Plants

INTRODUCTION

India has made great strides in industrialization during post-independence years. This has brought prosperity to a selected few and created concentrated centres of employment near the industrial complexes and big cities where economic activities have flourished. The material and energy resources have been diverted to feed these industrial and commercial 'gluttons'. The high technology industry could not penetrate the traditionally agricultural rural and remote areas where nearly 78% of India's population live. The rural population felt neglected and many of them are flocking to the new economy centres in search of their livelihood. The continuous migration of population and uneven distribution of wealth, material and energy resources have created many social and political problems which can lead to very serious consequences, if not further checked.

The solution of the problem lies in concerted efforts for the development, welfare and prosperity of the rural areas which in turn depend upon the availability of cheap and abundant energy for domestic cooking, heating and lighting community development, irrigation water, mechanized agriculture and small industries.

Transportation of commercial fuels like coal, diesel and petrol over long distances to remote and rural areas is very difficult and transmission and distribution of electricity through power grid to far-flung areas are prohibitively expensive. The farmers and craftsmen are not in a position to pay for high costs of energy, thus small-scale industries in these areas are reduced and there are limited irrigation possibilities in agriculture.

The modern high technologies do not adopt to rural ecology. An energy production technique based on renewable source and appropriate technology which can be taken care by trained local people for its operation and maintenance problems is ideally suited. Only appropriate technology involving local resources, manpower and raw materials and environment can ensure environmental harmony and sustainability of the development. The utilisation of energy in flowing water through micro hydro schemes is a good example of appropriate technology for rural development.

SMALL HYDEL PROJECTS IN INDIA

Hydro power, available free of cost in the form of flowing water, is an important, pollution free, renewable source of energy. The large potential of hydro energy available from flowing water in the rivers and irrigation canals, however, still remains untapped, since generating power on a small scale from low and ultra low head drops of 3 to 20 metres has not been an economic proposition because of the high civil and equipment costs as well as high operating cost to be incurred per unit of power generated. The situation is further deteriorated by very low load factors in rural areas.

China has developed 1200 MW of power by small hydel projects by creating water heads of about 5 m along the rivers. It is estimated that 5000 MW capacity of power can be created in India with the help of small hydro turbines of 200 kW to 1000 kW capacity each. M/s. Jyoti Limited, Baroda have the necessary know-how and capacity to manufacture these small hydro turbines, see Table 1. There are also other organisations showing interest in the manufacture of such small sets. Bharat Heavy Electricals Ltd. also has the necessary know-how which can be passed on to some ancillary unit for for quick manufacture of small hydro turbines without disturbing their main manufacturing plan for large sets.

Table 1 Manufacturing range of Jyoti Ltd. for small hydel sets

Type of Turbine	*Operating head range (m)*	*Output range (kW)*
Pelton	100 to 400	50 to 6000
Turgo Impulse	50 to 200	25 to 3000
Francis	20 to 200	50 to 3000
Propeller Kaplan	2 to 30	10 to 6000
Horizontal Tubular	2 to 25	10 to 6000

Due to civil costs of dams, water conduits and diversion channels, independent small hydel projects become very costly. It may cost Rs. 18000 per kW installed. Such projects should be planned on the basis of multiple purpose use of water resources such as irrigation and power generation. Even though these schemes are costlier when compared with the cost of unit power generation from major hydel projects, these are encouraging keeping long-term energy availability in view. Also hydropower is renewable in nature, once built, there is built-in protection against inflation of cost, water is available essentially free, it can be instantaneously switched on, it need not run continuously.

According to the size of hydroturbine, the small hydel projects may be classified as:

Micro Hydel Projects : 50 to 100 kW
Mini Hydel Projects : 100 to 1000 kW
Small Hydel Projects : 2 MW to 15 MW

Small hydel projects can be completed in four to five years, mini and micro projects in three years provided civil works of irrigation projects are already completed. Here water released for irrigation only is used and power is therefore, generated for 6 to 7 months a year.

In Maharashtra, 130 schemes of various catagories of small hydro projects of about 11000 kW capacity for a layout of Rs. 2000 millions have been planned. A separate circle has been created in UP State Electricity Board with headquarters at Nainital for small hydel projects. Sixteen micro hydel schemes are already functioning in eight hill districts of UP with total istalled capacity of 7730 kW. Four schemes are under construction for a total cost of Rs. 3.37 crores. The total installed capacity after completion of these schemes will be 8830 kW. Other 32 schemes of 20,000 kW are in the pipeline.

Other hilly states are also going for small and micro hydel schemes in a big way.

RIVER CURRENT ENERGY

Western Ghats of Maharashtra and regions of Uttar Pradesh, Himachal Pradesh, Jammu & Kashmir, Sikkim and North-Eastern States located in the Himalayas are suitable sites for small hydel projects where medium to high heads can be available. Equipment costs are low in the high head schemes. As we come down the hills, the available head decreases and equipment costs go up. For example, the cost of one hydro set of horizontal tubular design of 3.2 kW capacity operating at a head of 3 m may be Rs. 1.28 lacs, i.e., Rs. 40,000 per kW installed. The civil costs are site-specific but taking 30% of equipment cost, the total installed cost will be a high value of Rs. 52,000 per kW installed.

Most of our rivers flow through flat terrains which can be harnessed to produce power much in excess of 5000 MW as estimated above. Chinese model is difficult to develop due to different topography in India. Special 'zero' head turbines operating on kinetic energy of flowing water should be developed to harness huge river current energy. The civil costs of micro hydel power projects can be drastically cut by using floating type plants. There is no need of diversion channels, deep excavations and massive foundations. The floating plant consists of two floats with a vertical (or horizontal) runner in between which rotates by the velocity of water. The capacity developed is a function of the size of the runner and cube of water current speed. The rotational speed of runner is rather low and electric generator can be operated via a transmission gear. The floating plant moves up and down with water level and gives constant performance all the year round which is a big advantage over firmly installed water turbines especially during monsoons.

The current energy of flowing water which can be exploited by floating type power plant may be expressed as:

$$P = 1/2\, AV^3 E \text{ (kW)}$$

where P = Power generated

A = Embracing area of runner (m^2)

V = Water current speed (m/s)

E = Efficiency of runner depends upon the type of the runner and varies between 0.5 and 0.85.

Although hydraulic wheels will operate at low water velocities, it is unlikely to be an economic proposition in current speeds below 1 m/s (2 knots).

PONCELET WATER WHEEL

In ancient times, floating mills were used in Germany on large rivers such as the Rhine, the Weser and the Elbe to drive electric generators and machine tools. A vertical undershot water wheel consisting of a horizontal shaft and radial blades fixed to it, was fitted on the side of a barge which was either moored to the bank or firmly anchored in the stream. A smaller barge was used to support the other end of the shaft. The hulls of the barges were built with local materials. The efficiency realised were very low, in the range of 20 to 30% because of bad blade geometry (straight radial blades).

Poncelet water wheel is an improvement on the straight blades of undershot water wheel, which have suitable curvature. Water strikes the vanes practically without shock (inlet vane angle 15°) and drives it by impulse. Water is discharged from the blades almost vertically downwards. The efficiencies achieved are 55 to 65%.

The main drawback of such machines is low power output for a given size and weight because at any instant only a small part of it is actually in the water being driven by the current while the rest of the machine is idle.

DARRIEUS TURBINE ROTOR

Darrieus turbine rotor is similar to modern vertical-axis wind turbine which was first introduced in France by G.J.M. Darrieus in 1920s. Its configuration consists of four hydrofoil blades which rotate at much higher speed than the water current speed. The speed of rotation is primarily dependent upon the machine overall dimensions. It has relatively low starting torque. Because of simplicity of the blade design and because they are relatively thin blade, fabrication costs are low. They require no pitch control for synchronous applications.

Two prototypes have been developed by Intermediate Technology Development Group, London and are shown in Fig. 1 and 2. The blades can be made out of timber or fabricated from steel, ferrocement or glass fibre. The expected performance from these models are plotted in Fig. 4.

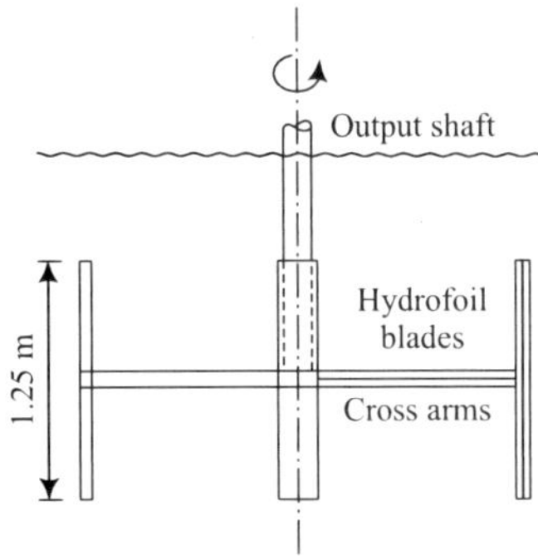

Fig. 1 Vertical axis rotor 13.5 rpm in 1 m/s current power take-off above surface swept area 3.75 m^2

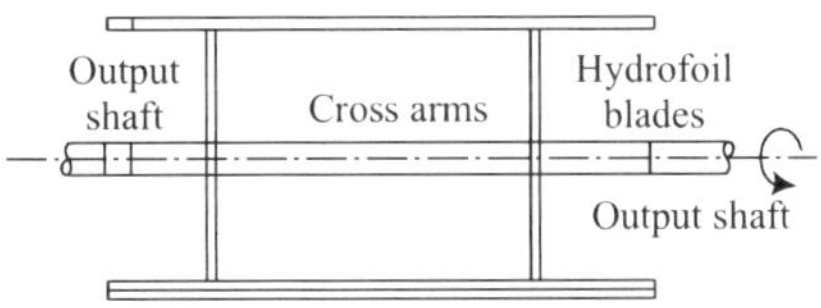

Fig. 2 Horizontal axis rotor 32 rpm in 1 m/s current power take-offs under water swept area 3.75 m^2

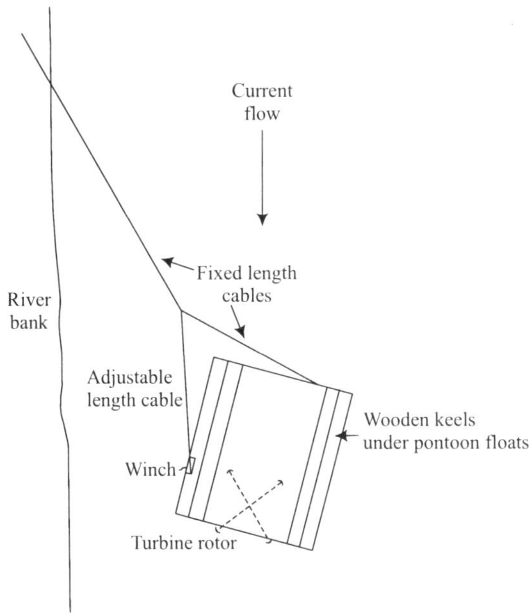

Fig. 3 Mooring system for first prototype

INSTALLATION

Floating type micro hydro power plants are suitable for use in situations where there is no 'head' of water available and is designed to extract kinetic energy from a river or canal current. Therefore, appropriate sites are along the banks of large rivers or on irrigation canals.

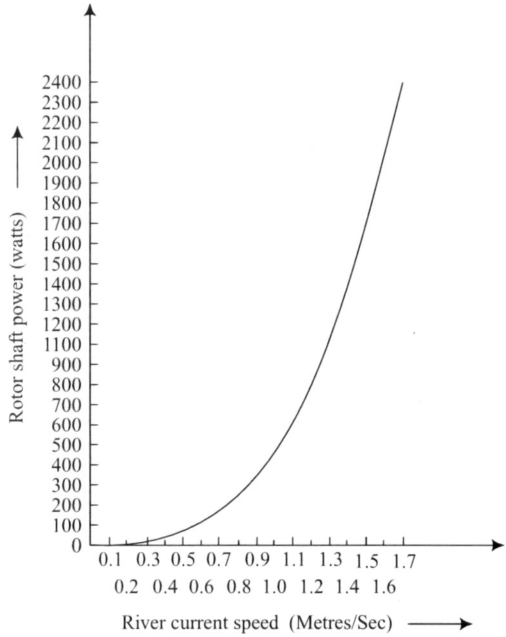

Fig. 4 Graph showing how turbine rotor power varies with river current speed for a turbine with swept areas 4 m^2 and rotor efficiency 25%

As the energy flux depends upon the cube of water velocity, the latter must be as large as possible. The depth of water required for a 4 m^2 runner is at least 2.5 m.

The Darrieus turbine rotor is a low speed machine. It is a low solidity device which is completely submerged during operation and this, together with its high efficiency (above 50%), means that a rotor of moderate size and modest materials content can extract useful amounts of energy from current speeds as low as 1 m/s (2 knots).

Irrigation canals are expected to be an important area of application for floating type micro hydro power plants. In a large canal, the turbine rotor could be mounted on a pontoon as in a river situation, whereas on a small canal it might be cheaper to mount the rotor on a bridge or beam over the canal. The power output from the rotor could be increased by shaping the canal banks to form a venturi to increase the water velocity through the rotor.

The mooring details and site requirements are shown in Figs. 3 and 5 respectively. The mooring cable arrangement protects the rotor against submerged or semi-submerged objects striking the blades.

The pontoon which carries the turbine, generator and the transmission system, can float on two rows of barges or oil drums under which wooden keels are affixed. The mooring cable can be adjusted with a winch to ensure sufficient angle

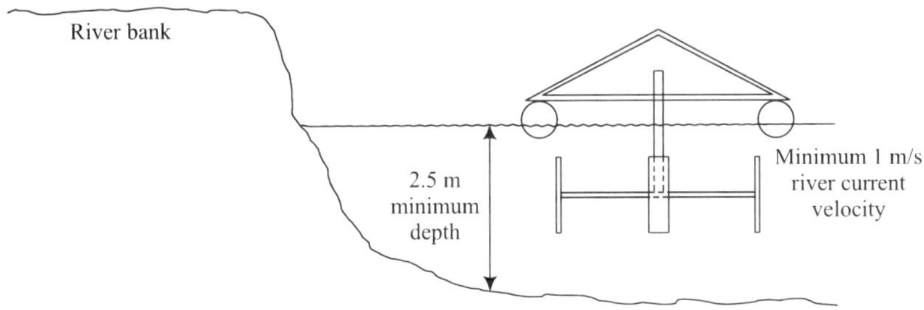

Fig. 5 Site requirement for turbine with 4 m^2 swept area

of the keels to the current direction so that water side thrust on the keels keeps the pontoon at a distance from the bank.

AUXILIARIES

The rotational speed of Darrieus turbine is very low: 13.5 rpm for vertical axis and 32 rpm for horizontal axis turbines as shown in Figs. 1 and 2. For Poncelet turbine, the peripheral velocity of runner is about half of water current speed. Therefore, multi-stage belt drives or gear boxes are needed to increase the speed of the generator shaft to about 950 rpm. Permanent magnet alternators or induction generators are preferred over synchronous generators. To reduce equipment costs, the conventional mechanical or hydromechanical speed governor should be dispensed with. The inertia of rotor being small, there is a problem of instability in power supply quality.

The above problems can be solved by use of a flywheel which can ensure stable power supply and can shave-off power demand fluctuations. A flywheel designed for this purpose can be accelerated to extremely high speeds without any risk of breaking apart. It can be made from concentric rings of quartz fibre as windings, the rings being fitted over one another and close gaps filled with an elastic substance to keep the plies of the rims together. The flywheel is coupled to a generator and enclosed in a sealed evacuated casing to reduce windage losses. The device operates as a generator when power demand on the system grows, and as an electric motor when it is time to accumulate energy. According to some calculations, the cost of 1 kW will be less than with pumped hydroelectric storage. A project has been described contemplating a 5m diameter flywheel of 1.96 MN weight that could be charged to 20 MWh. The operating speed is 3500 rpm.

The mechanical energy storage system as described above can fit into appropriate technology level suitable for rural environment. The load governors based on electronic circuitry or microprocessor technology may be able to solve the problem to some extent but may not be easily acceptable in rural areas because of higher levels of technology involved. Further, these load governors do not operate happily against unstable operations of turbine rotor. Load governors invariably require an auxiliary load or another hybrid system to store surplus electric power generated by hydel power plant during low demand period. This is not always a good decision thermodynamically as electrical energy is the highest form of energy and hybrid systems help to degrade the energy form and in addition there are heavy conversion losses.

CONCLUSION

Floating type micro hydro power plants can be a very attractive proposition to harness enormous amounts of hydroenergy available as river current energy in the flat countryside. Civil costs are reduced and cheap and reliable equipment can be developed for tapping the flow energy in the irrigation canals by constructing the banks into venturi shapes. Research and development efforts are needed to develop efficient turbine rotors working on the principle of wind turbines or old time water wheels followed by intensive field trials. The problem of stability and energy storage to improve load factors can be solved by the development of suitable flywheels. The developmental efforts must be constrained keeping in view the principles of appropriate technology suitable for harmonious development of rural areas involving local people, local materials and local resources.

REFERENCES

1. Peter German, 'The Development of a Turbine for Tapping River Current Energy', Intermediate Energy Technology Development Group, London.
2. 'Upgrading and Further Development of Traditional Devices for the Utilisation of Hydroenergy', Bremen Overseas Research and Development Association, Bremen.
3. Jagdish Lal, 'Hydraulic Machines', Metropolitan Book Co. Pvt. Ltd., Delhi.
4. N.P. Cheremisinoff, 'Fundamentals of Wind Energy', Ann Abor Science Publishers Inc., Michigan, 1978.
5. P.D. Mohindra, 'Current from Currents' *The Tribune*, 14th June. 1982.
6. Brij Bhardwaj, 'Surplus Power by 1986 Envisaged' *Hindustan Times* dated 5th April, 1984.
7. 'UP submits Five more Micro Hydel Schemes,' *Hindustan Times* dated 30th Sept., 83.
8. Venikov & Putyatin, 'Introduction to Energy Technology' Mir Publishers, Moscow, 1981.

CASE **II**

An Alpine Hydro Village

INTRODUCTION

India is a country of villages. These villages vary in size and population and are scattered all over the country. Some of these are located on the hills and in dense forests and can be reached only on foot or by pony. They have no road connectivity to other villages, towns and cities. The population of these villages may be one hundred to two hundred people. Most of them earn their living by cultivating the land, growing fruits and some of them may be craftsmen who earn their living by weaving woollen carpets, by making earthen pots, by weaving baskets and even by making beautiful carved articles from wood, brass and other local materials. Some of them keep cattles like cows, sheep and goats and make their living on animal yields. In a nutshell, life is hard and hilly villages are very low income areas.

The main wealth of hilly areas is vegetation varying according to temperature, soil and rain and mainly depends upon the altitude. Near the base of the mountains, the vegetation will be thick evergreen forests while the top of the mountains will have no vegetation, but will be covered with eternal snows. The vegetation belts of the Himalayas are given in Table 1.

Table 1 Vegetation belts of the Himalayas

Altitude (m)	*Description*	*Trees*	*Fruits*	*Crops*
0-750	Tarai	Sal, Bamboo	Tropical Fruits	Barley, Rice, Tea, Potatoes
750-1000	Deciduous Forests	Oak, Horse-Chestnut	Temperate Fruits	Rice
1000-1600	Mixed Forests	Oak, Beech	Walnuts	Tea
1600-3300	Coniferous Forests	Pine, Chir, Fir, Deodar, Cedar, Spruce	—	—
3300	Timber	Line		
3300-4200	Alpine Grasslands	—	—	—
4200	Snow	Line		
4200-8500	Eternal Snow	—	—	—

The valleys are surrounded by the mountains and are protected from the cold winds blowing on the mountains. The climate is warm and the soil is more fertile than that of the mountains. In almost all valleys, there are rivers and sometime lakes. Good crops especially fruits like apples, pears, cherries, etc., can be grown in the valleys.

India has made great strides in industrialisation during post-independence years. This has brought prosperity to a selected few and created concentrated centres of employment near the industrial complexes and big cities where economic activities have flourished. The material and energy resources have been diverted to feed these industrial and commercial 'gluttons'. The high technology industry could not penetrate the hilly areas. The population of these areas felt neglected and many of them are flocking to the economy centres in search of their livelihood. The continuous migration of population and uneven distribution of wealth, material and energy resources have created many social and political problems which can lead to very serious consequences, if not further checked.

NEED FOR CHANGE

The continuous flux of population to new economy centres has created the problems of mushrooming of slums and destruction of social life; the introduction of high technology will give birth to widespread unemployment. These problems can explode with very serious consequences. These can be checked only by adoption of Gandhian model of socio-economic development with village as the unit. Each unit must be self-sufficient economically and the ratio of exports to imports of goods into a village must be more or equal to one. The development, welfare and prosperity of hilly villages depend upon the availability of cheap and abundant energy for domestic cooking, heating and lighting, community development, irrigation water, mechanized agriculture and small industrial parks based on local materials.

Transportation of commercial fuels like coal, diesel and petrol over long distances to hilly areas is very difficult and transmission and distribution of electricity through power grids to far flung areas are prohibitively expensive. The farmers and craftsmen are not in a position to pay for high costs of energy, thus small-scale industries in these areas are reduced and there are limited irrigation possibilities in agriculture.

APPROPRIATE TECHNOLOGY

The modern high technologies do not adopt to rural ecology. Only appropriate technology involving local resources, manpower and raw materials can ensure environmental harmony and sustainability of the development. However, there should be provision for upgrading of technology in the future. The wealth of the hills can be increased by inputting ample energy and technology to increase agricultural, forest and animal yields. Appropriate technology can be worked out by a technical institute specially established for the development of hilly areas. An energy production technique based on local resources and appropriate technology which can be taken care by trained local people for its design, manufacture, installation, operation and maintenance is ideally suited. The utilization of energy flowing water through small hydel schemes is a good example of appropriate technology for the development of hilly areas. An alpine hydro village should meet most of its energy requirements from small hydel schemes. One such model is described in the following paragraphs.

The utilisation of hydroenergy to operate agriculture and industrial devices is one of the oldest and widespread techniques of human achievements. Hydraulic rams have been used to raise water for lift irrigation and drinking purposes. Water mills of various types have been used to generate mechanical power to drive irrigation pumps, machine tools and small electric generators. The Himalaya Mill is well known in the mountain areas from Afghanistan to Burma which is mostly used for grinding grains. The present model of alpine hydro village is based on these devices and are briefly described before discussing the hydro model proper.

HYDRAULIC RAM

Hydraulic ram is a contrivance to raise a part of large amount of water available at some height, to a greater height. This can be employed in hilly areas where some natural source of water like a spring or a stream is available at some altitude. Work done by a large quantity of water in falling through a small height is used to raise a small part of it to a greater height. Action of water hammer makes it feasible.

No external power is, therefore, required to work this machine. Other attractions are: negligible amount of maintenance and supervision costs, continuous operation, high efficiency, quiet operation and possibility of automatic adjustment of water supply. The liftable volume of water diminishes asymptotically with lifting height. In case of medium lifting heights, the hydraulic ram operates with an efficiency absolutely comparable to a piston pump of the same performance.

As the hydraulic ram does not need a driving unit, it is to be considered ideal for lift irrigation and supply of drinking water for hilly areas, where the supply in fossil energy carriers or electricity is problematic. The hydraulic ram operates with water streams of 1 to 40 m^3/s and fall heads of 1.5 to 30 m and with lifting heights upto 300 m. There are a number of different types which can be distinguished by the operating of oscillating valves.

Hydraulic ram is a simple and rugged device for operating irrigation schemes. This can be easily assembled from local materials. The proposed technical institute for the development of hilly areas has to put in some developmental work to improve the performance and adoption of hydraulic ram to irrigation.

HIMALAYA MILL

It is a vertical shaft, horizontal runner hydraulic turbine used to drive grain mills. The wooden radial blades are attached to a boss fitted into a wooden or steel shaft. The water is fed through open wooden channel. It produces about 0.5 hp with a water head of 1m. There are about 10,000 such mills scattered in the hilly areas from Afghanistan to Burma producing power equivalent of about 45000 kWh daily.

The horizontal water wheel can be improved in performance to raise its efficiency substantially so that it can be adopted in a small-scale industry park as a prime-mover for flour mills, threshers, rice-husking machines, vertical lathe, circular saw, trip-hammer, small electrical generator, carpet loom, etc. The vertical lathe can be used by the craftsmen to produce wooden vessels and trip hammers for embossed copper vessels. The sawmill can be used by the carpenters. All these machines can be employed together to meet the local needs to produce domestic and agricultural implements and packing boxes for the export of fruits and other products.

The Himalaya Mill can be developed into an efficient and modern Pelton turbine using metallic buckets fitted to a vertical shaft supported on roller bearings. The water can be directed onto the buckets, through adjustable nozzle from a piping system collecting water from natural streams existing abundantly in the hilly areas. The stopping of turbine can be achieved through manual brakes and load/speed regulation by the nozzle. The design documentation can be provided by the proposed technical institute for the development of hilly areas and machine can be manufactured in the small-scale industry park. The output of the machines can be standardised at 5 to 10 kW.

The tail race water can be recirculated by lifting it with a hydraulic ram for irrigation purposes.

FLOATING MILL

Poncelet water wheel of a Darrius Turbine rotor can be submerged in a river or water stream available in the hilly areas to tap current energy from water of high speeds. The mechanical energy thus obtained is used for driving irrigation pumps, small electric generators and small compressors for cold storages and ice mills. The small cold storages are necessary in the hilly areas to store potatoes and other perishable goods to get better returns. Ice mill can be used to chill and store milk before it is sent to bigger collection centres.

The electricity generated is required to light the houses, schools, community centres and to run TV sets and other domestic devices. All these measures can improve the wealth of the village multifold and raise the standard of living of the population which will arrest their migration to big cities.

The blades of the water mills can be made from fabricated steel, ferrocement, glass fibre or timber depending upon the availability of the local materials and skills. The power output depends upon the swept area and cube of water velocity and very large energy can be tapped from hill streams flowing at high speeds.

The turbine rotor can be mounted on a pontoon or may be hung from a bridge or a beam over the stream. The pontoon can float on two rows of barges or empty oil drums under which wooden keels are fixed. The mooring cable can be adjusted with a winch to ensure sufficient angle of the keels to the current direction, so that water side thrust on the keel keeps the pontoon at a distance from the bank. A turbine with swept area of 4 m^2 and rotor efficiency of 25% can give an output of 8 kW at a water current speed of only 2 m/s. These turbines can be adopted to run small cold storages and ice mills, irrigation pumps and to produce all the electricity requirements of the village.

HYDRO VILLAGE

A typical layout of an alpine hydro village is shown in Fig. 1. The capacity and number of each type of hydro machines will depend upon the size of the village and characteristics of available water stream. The following units have been included in the scheme.

1. A small cold storage and ice mill run by a floating water mill. The capacity of the plant can be around 10 kW. Such cold storages can be built along the flow of water stream to store potatoes, fruits and other perishable products. This will help to avoid spoilage of such products and realize better returns by disposing of uniformly over the year. Similarly, the ice mills attached to the cold storage can be used to chill and store the milk cooperatively and can be drawn for disposal to collection centres and self use efficiently.
2. Electrical generator run by a floating water mill. The capacity will match the village requirement of electricity for lighting, heating, cooking and

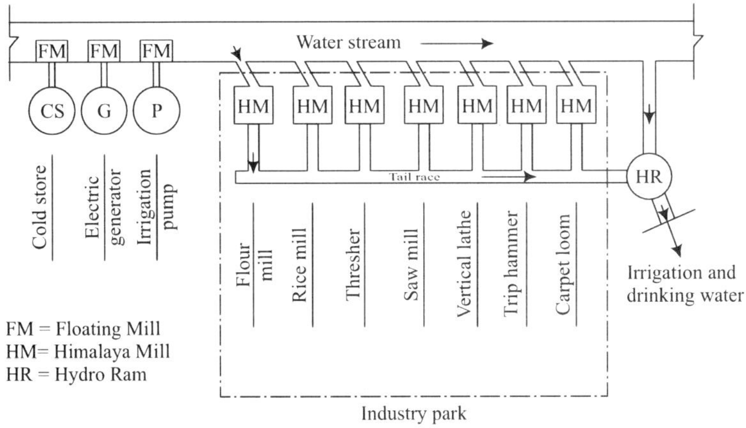

Fig. 1 Hydro village

running of radios, TV sets and other small electrical gadgets in the houses, schools, dispensaries and community centres.

3. Irrigation pump run by a floating water mill. The capacity will suit the irrigation water requirements.
4. Industry park consisting of flour mills, rice mills, threshers, saw mills, vertical lathes, trip hammers, carpet looms, etc., run by improved Himalaya mills individually. The capacities can be standardised at 5 to 10 kW each. The industry park will meet all the domestic and agricultural requirements and make the village independent of imports of implements and spare parts. This will help to produce the goods like wooden, brass and other vessels, toys, beauty carvings, woollen carpets for exports with mechanical aids. All packing boxes for export of forest, agricultural and animal yields will be prepared in the industry park.
5. Hydraulic ram to lift the used water from tail race for irrigation and domestic water consumption.

CONCLUSION

The proposed alpine hydro village can meet the productivity requirements for self-sufficiency as per Gandhian Model. A technical institute for the development of hilly areas should be set up who can undertake research and development for the improvement and adoption of hydraulic machines for a hydro village. It should also plan and implement such schemes and monitor the results. Such hydro villages can go a long way in bringing prosperity and self-sufficiency to these villages. The migration of population can be arrested and the county can be prepared to absorb high technology in the required sectors.

REFERENCES

1. Singal, R.K., ‘Floating Type Micro Hydro Power Plants’. *Electrical India*, Bombay, 30th September 1984.
2. Singal, R.K., ‘Non-Conventional Energy for Lift Irrigation’ *Electrical India*, Bombay, March, 1985.
3. ‘Upgrading and Further Development of Traditional Devices for the Utilisation of Hydroenergy’, Bremen Overseas Research and Development Association, Bremen.

CASE III

Operation and Maintenance of Centrifugal Pumps for Energy Conservation

In many industries, pumps may be forming more than 40% of the auxiliary equipment and if proper design and operation of the pumping equipment can help to conserve even a few per cent of the huge power consumed by them, it will serve our country well. Here we will examine various sources of power saving which can be achieved in pumping installations by adopting proper approach to their operation and maintenance.

Never before in the history of industrial society we have faced as pressing a need to conserve energy as now. The high cost of energy and scarcity of fuel such as oil or natural gas, or even coal have become a hard fact of life. In addition, new constraints are being introduced on the use of power from the point of view of ecology.

Therefore, it is imperative to examine all energy-consuming processes with a view to improving their overall efficiency. And since every industrial process which underlines our modern civilization involves the transfer of liquids from one level of pressure or static energy to another, pumps have become an essential part of all industrial processes, and, in turn, major consumers of energy themselves.

Some important applications of centrifugal pumps are given in Annexure-I showing the widespread use of pumps in various sectors of economy. In many industries, pumps may be forming more than 40% of the auxiliary equipment and if proper design and operation of the pumping equipment can help to conserve even a few per cent of the huge power consumed by them, it will serve our country well.

PROPER IMPELLER DIAMETER

A centrifugal pump is a velocity machine. The peripheral speed of the impeller depends directly upon its diameter. Therefore, the pump capacity which is a direct function of the velocities, will vary directly as the impeller diameter, the total head, which is a function of the square of the peripheral speed, will vary as the square of the impeller diameter.

Finally, since power consumption varies as the product of the head and the capacity, the power will vary as the cube of the impeller diameter. Therefore, it is very important to run a pump with proper size of the impeller. However, one of the greatest sources of power wastage is the practice of oversizing a pump by selecting operating conditions with excessive margins in both capacity and total head.

The pump is then operated with considerable throttling to limit its delivery to the desired value thereby wasting power by as much as 5, 10 or even 15%. The point in question can be illustrated by an actual example given in Annexure-II. There is a saving in power consumption of 11.5% by using an impeller of 355 tip diameter in place of 375 mm diameter. The power savings may jump to 18% initially when the piping is new. There is additional saving in capital expenditure due to smaller size of motor required.

It is true that some margins should always be included, mainly to provide against the wear of internal clearances which will, with time, reduce the effective capacity of the pump. However, reasonable restraints must be exercised to avoid incorporating excessive safety margins into the rated capacity conditions of service. It will be a safe practice to design a pump with normal margin of 10% either on capacity or head (not both) and order a pump with impeller one size smaller.

On a future date, when the internal clearances have increased due to corrosive and erosive action of the liquid, or additional demand is created, the bigger impeller can be procured and installed into the existing casing to cover up the additional demand in pump capacity.

On an existing installation, run a performance test and plot the true system – head curve. A reasonable margin can be selected and three choices become available to the operator.

- Trim the existing impeller to meet the more realistic condition of service required for the installation. The diameter ratio can be approximated from the square root of the two pressures. After the impeller is cut the vane tips should be filed very carefully to approximate the finish of the vanes as they appear new.
- A replacement impeller with the necessary reduced diameter may be ordered from the pump manufacturer. The original impeller can be stored for future use for meeting the loss of pump capacity due to increased friction losses.
- A replacement narrower impeller may be ordered from the pump manufacturer and the original impeller may be stored for possible future use. A narrower impeller will have its best efficiency at a lower capacity than the normal width impeller.

VARIABLE-SPEED DEVICE

Since it is the peripheral speed of impeller that determines the total head and the capacity of the pump, obviously it is immaterial whether the peripheral speed is changed by cutting down the impeller diameter or the pump speed. When the speed is changed, the capacity for any given point on the pump characteristic curve varies directly as the speed, and at the same time, the head varies as the square of the speed while the power varies as the cube of the speed.

A magnetic drive or a hydraulic coupling can be interposed between the pump and the electric motor to meet the exact required conditions of service without throttling by reducing the pump operating speed.

The method of calculations of speed change is illustrated in ***Annexure-III***. The potential power savings are huge. If the number of hours of operation at various expected pump capacities are known, the yearly savings in kWh and energy cost can be calculated to justify the installation of variable speed device. If techno-economically it is not feasible to install a variable speed device, use smaller impeller as discussed in para 2.

PARALLEL OPERATION

In most of the installations, two 50% capacity pumps are used which operate in parallel to deliver the required flow under full load conditions. However, the flow requirements vary widely and important power savings are possible through improved operating practices. It is clear from ***Annexure-IV*** that very important power savings are possible if one pump can be shut down when the flow requirements are 50% or less.

The evaluation of any energy conservation programme consists of the comparison between the amount of energy saved and the cost of the implementation of the programme. The remarkable fact is that the cost of implementation in this case is exactly zero. As a bonus, the life of the pump improves as pump when operating more nearly to their best efficiency point have maximum life.

MAINTENANCE OF INTERNAL CLEARANCES

The leakage losses in a pump increase due to wear of internal clearances, thereby affecting the HQ curve and power input. The rate of wear depends upon differential pressure across the pump, type of fluid handled (corrosive and abrasive) severity of service, material of pump internal parts, etc. The effect of leakage losses on power consumption depends upon the specific speed of the pump i.e., lower the specific speed, the head per stage is higher and therefore the wear is increased due to higher differential pressure.

Therefore, there is a need to attend to restoration of internal clearances specifically for pump with lower specific speed. The effect of wear on the deterioration of HQ curves is shown in Fig. 1.

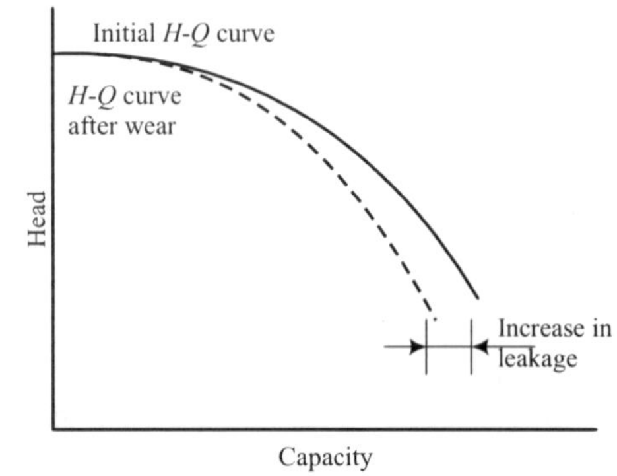

Fig. 1 Effect of internal wear on H-Q curve of C F pumps

The pressure drop across all the internal joints is of the same order of magnitude when expressed in terms of pressure drop across units of linear dimensions, all running joints will wear at about the same rate. It is thus possible to predict when internal clearances should be renewed without dismantling the pump.

The leakage through a balancing device is returned to the pump suction via an external pipe, in which a flow measuring orifice can be located. This leakage can therefore be readily measured. The measurement can be used to monitor the progress of the wear experienced by the pump.

The relation between leakage losses in percentage of power output and specific speed is shown in Fig. 2. The leakage losses when the pump is new are of the order of 1.2 per cent for pumps of high specific speeds of 2200-2500. Thus, when the internal clearances have increased to the point that this leakage has doubled, we can regain approximately 1.2 per cent in power savings by restoring the pump clearances.

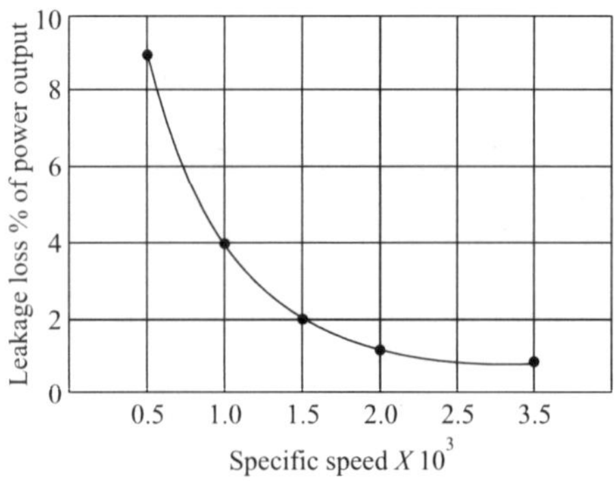

Fig. 2 Leakage losses for double suction pumps

On the other hand, for pump with low specific speeds of the order of 700-800, the leakage losses are of the order of 5 per cent. If the clearances are restored after the leakage losses doubled, we can count 5 per cent power savings. Therefore, restoring clearances of the lower specific speed-type pumps gives us greater returns in terms of the reduction of leakage losses.

It may well be, then, that economic conditions may permit us to let clearances of higher specific speed pumps increase beyond values double those of a new pump while making it attractive to review internal clearances of low specific speed pumps before they have doubled. The final decision, at any rate, must be based on comparing the costs of an overhaul with the value of the potential power savings, considering actual energy costs.

MECHANICAL PROBLEMS

In addition to hydraulic or system design, a number of mechanical problems must be attended to save excessive power consumption in the operation of pumps.

- Rubbing between the rotating and stationary parts of a pump, for whatever cause, will of course increase the power consumption. This rubbing may be caused by misalignment of pump and driver, by a bent shaft or by incorrect reassembly of a pump during maintenance and replacement of internal parts.
- Excessive wear of parts at the internal running clearances will generally lead to increased power consumptions. As already discussed, increased clearances cause increased internal leakage and the impeller is forced to handle a much higher flow than is being discharged by the pump into the system.
- Improper conditions at the stuffing boxes will frequently increase power consumption. This is particularly true for small pumps for which the mechanical losses at the stuffing boxes represent a significant portion of the total power consumption. The packing may have been improperly installed or it may be inadequate for the actual operating conditions; the glands may have been tightened excessively. Anyone of these problems will increase the friction losses at the packing.
- If a significant foreign matter becomes lodged in the impeller water ways, the pump efficiency will suffer and there will be increased power consumption.

CONCLUSION

The operator of a pump must analyze each of the possibilities, starting with the most probable cause, and to correct the problem that leads to increased power consumption. Potential of energy conservation is huge. Tap it.

Annexure-I

APPLICATIONS OF CENTRIFUGAL PUMPS

Rural Sector

- Agriculture: Irrigation, farming, insecticides and fertilizer spraying, forestry, fish farms and hatcheries
- Water Supply: Drinking and waste water
- Cottage & Rural Industries

Transport Sector

- Harbours, docks and canals.
- Road, rail and air tranport Tranportation equipment manufacture, e.g. car, lorry, bus, aircraft, rolling stock and naval.
- Marine Commercial ship-building, and ship owners.
- Mining & Quarrying Coal mining, metallic and non-metallic ores, clay, salt, sand, gravel, etc.

Construction

- Pipeline construction
- Civil engineering works e.g. harbours and dry docks, road, bridges, etc.

Service Industries & Services

- Heating, ventilating and air-conditioning, e.g., hospitals, schools, offices, etc.
- Domestic water supply, e.g. domestic booster-sets and domestic hot water circulations.
- Fire-fighting and fire prevention, e.g. sprinkler system, etc.
- Armed forces, atomic energy establishments, research establishments postal services and communications and other miscellaneous local government services.
- Laundries, dry cleaners and dryers, sports establishments.
- Commercial construction services, hotels, hospitals, schools, offices.

Power Generation and Distribution

- Gas: Processing and distribution.
- Electricity: Electricity generation and transmission.

- Water supply: Municipalities and other agencies.
- Sewage: Sewage treatment and disposal.

Oil Explorations and Production

- Oil and natural gas exploration and production.
- Petroleum refinery plants
- Petroleum distribution centres
- Coal, coke and tar processing

Engineering Industries

- Mechanical machinery and component manufacturers, e.g. machine tools, pump, compressors, food machinery, heaters, coolers, fasteners, instruments, etc.
- Boiler manufacturers.
- Dish and bottle washing plant manufacturers.
- Metal parts washing and spray painting plant manufacturers.
- Filtration plant manufacturers
- Commercial industrial refrigeration.
- Electrical motors, generator, radio, television, radar, telephony equipment, wires and cables, other domestic electrical equipment.
- Switchgear manufacturers.

Food and Beverages

- Cereal processing, bakeries and confectionery products, canneries; fruit and vegetable processing and products, animal and marine fats and oils; dairy products, e.g. milk, cheese, butter, ice creams, meat processing, e.g. slaughter houses and bacon factories.
- Distilleries, breweries, soft drink manufacturers.
- Cane and beat sugar processing.
- Tobacco.

Process Industries

- Petrochemicals, plastic and synthetic rubber manufacturer.
- Chemical manufacturers, e.g. fertilizers, inorganic chemicals, organic chemical (acids, solvents, ammonia, chlorine, etc.) inks, paints, photographic chemicals, polishers.
- Water treatment plant manufacturers and contractors.
- Pharmaceuticals and cosmetics manufacturers.

- Paper, pulp and printing manufacturers and contractors, printing, publishing and book binding.
- Textiles: Synthetic fibres manufacturers.
- Consumer goods: Textiles, tanneries and leather processing, furs, leather goods manufacturers, clothing and footwear, plastic consumer goods manufacturers.

Industrial Materials

- Primary metal production, foundries and rolling mill equipment.
- Electropolishing and plating specialists and plant manufacturs.
- Bricks, fire-clay and pottery, glass and glass-container manufacturs, cement readymix concrete.
- Timber, furniture and bedding, other wood and cork manufacturs.

Annexure-II

SELECTION OF A CENTRIFUGAL PUMP

- **Design Conditions**

Max. desired capacity	= 200 lps
Static head	= 35 m
Total friction losses assuming 15 years old pipe	= 18 m
Total head required	= 53 m
Margin on capacity at the rate of 10%	= 20 lps
Total design capacity	= 220 lps
Margin on total head	= 7 m
Total design head	= 60 m

- **Pump Selection I**

Impeller diameter	= 375 mm
Pump power required at 220 lps	= 130 kW
Motor capacity	= 150 kW
Power consumption at rated capacity and head (achieved by throttling)	= 123 kW

- **Pump Selection-II (Alternate)**

Impeller diameter	= 355 mm
Capacity at intersection with head capacity curve	= 210 lps
Margin on capacity	= 5%
Power consumption at intersection at 200 lps capacity	= 108 kW
Motor capacity	= 110 kW
Saving in power consumption	= 11.5%

Annexure-III

CALCULATION OF VARIABLE PUMP SPEED

The head-capacity curve of pump at 180 rpm is known. We have to determine the pump speed to deliver 150 lpm at a total head of 45 m. This point does not lie on the available *H-Q* curve.

- Select an arbitary capacity greater than 150 lpm, such that it will be located on the parabola defined by the affinity laws:

$$\frac{Q_3}{Q_2} = \frac{n_3}{n_2};\ \frac{H_3}{H_2} = \left(\frac{n_3}{n_2}\right)^2 = \left(\frac{Q_3}{Q_2}\right)^2$$

  Assume $Q_3 = 185$ lpm

$$H_3 = H_2\left(\frac{Q_3}{Q_2}\right)^2 = 45\left(\frac{185}{150}\right)^2 = 68.45 \text{ m}$$

- Draw the portion of the parabola defining the affinity laws between 150 lpm, 45 m and 185 lpm and 68.45 m. It can be assumed that this is essentially a straight line.
- The intersection of this straight line with the head-capacity curve of the pump at 1800 rpm corresponds to 180 lpm and 65 m. We can now determine the speed required to meet the desired conditions of 150 lpm and 45 m.

$$n_2 = 1800\sqrt{\frac{45}{65}} = 1497 \text{ rpm (say 1500 rpm)}$$

- If variable speed device is not available or cannot be justified on the basis of energy savings, the use of smaller or narrower may be used.

  In the above example, say the original impeller has a diameter of 375 mm.

$$D_2 = 375\sqrt{\frac{45}{65}} = 312 \text{ mm (say 315 mm i.e. the next size of impeller available with the manufacturer)}$$

- The power input to the variable speed device can be calculated as:

$$P_2 = P_1\left(\frac{n_2}{n_1}\right)^3$$

$$\frac{P_2}{P_1} = \left(\frac{1500}{1800}\right)^3 = 0.578 \text{ m.}$$

i.e. a saving of 42%

Annexure-IV

PARALLEL PUMP OPERATION

Full Load Conditions

Flow rate	= 240 lps
Static head	= 6 m
Friction loss in the system	= 9 m
Total head	= 15 m
Mode of operation	= Constant speed

Pump Selection

No. of pumps	= 2
Capacity of each pump	= 120 lps
Total head	= 15 m
Power consumption of each pump	= 18 kW
Total power consumption of the installation	= 36 kW

Part Load Operating Conditions

Flow rate required	= 120 lps
Total head	= 15 m

Both Pump Operating

Flow rate of each pump	= 60 lps
Head	= 20 m
Power consumption	= 15.6 kW
Total power consumption	= 31.2 kW

Single Pump Operating

It is possible to shut down one pump and meet the requirements with throttle considerably reduced.

Flow rate	= 120 lps
Head	= 15 m
Power consumption	= 18 kW

Power Savings

Savings resulting from single pump operation

= (31.2 – 18) = 13.2 kW

$$= \frac{13.2}{31.2} \times 100 = 42.3\%$$

Yearly operating time when flows required are 50% or less = 20%

Yearly savings = 8%.

CASE **IV**

Hydraulic Design of a Boiler Feed Pump to Ensure Stable Operation at Reduced Flows

The boiler feed pumps for industrial and power station boilers have to operate often at reduced capacities to meet the changing demand of steam and electricity. The operation of centrifugal pumps at reduced capacities lead to a number of unfavourable results seriously affecting the pump reliability. Some of these, such as internal recirculation of flow inside the pump have been recently studied. Here we will discuss these unfavourable results and analyse various design factors which can control unstable operation of the pumps at reduced flows. The commissioning problems of boiler feed at Rajasthan Atomic Power plant at Kota and modifications carried out in the light of the above studies are described below.

INTRODUCTION

The demand curve of total electrical power for a city or a region varies widely over a day, a month and a year. The fluctuations in the demand are expected to vary much more with further industrialization and improvement in the living standards as less and less industries might be working at night with longer stoppages. The fluctuations in hourly demand of electricity may be as large as 30% or even more of the peak demand. It is very difficult to force a change in the energy consumption pattern even by differential tariff of power. Therefore, the electrical power supply system must be designed to cope with the demand schedule.

It was estimated that by the year 2000 AD, more than 95% of electrical power will be produced by thermal and nuclear power plants. These plants are high temperature systems and most suited to operate at constant loads. These must be especially designed to meet wide load fluctuations.

Boiler feed pump is the most important link between the boiler house/nuclear reactor and the turbine hall. The load changes or even short shut-downs of the power plants will necessitate the feed pump operation at reduced flows. The industrial boilers generating steam for process heating and/or power generation for captive uses have to operate frequently at variable capacities and very often at 30% to 50% or even at 10% of rated value. Operation of centrifugal pumps at reduced capacities leads to a number of unfavourable results which may take place separately or simultaneously. The pump operates at the best efficiency point, the increased thrusts overload the bearings, the temperature rise through the pump increases, probability of flow separation and cavitation damage increases, hydraulic surges due to unstable head-capacity characteristics or internal recirculation of flow or steam flashing may cause accelerated cavitation, vibrations, noise. All these factors causing poor operation and premature impeller wear and deterioration must be avoided. Here we will discuss these problems in some details with reference to the experiences of nuclear power plants. The pump designers must take care of these phenomena to ensure stable operation of boiler feed pumps at reduced flows.

DESIGN PARAMETERS

The boiler feed service presents a number of problems not encountered in any other field of application. Multi-stage, single-suction, opposed impellers, horizontally split casing centrifugal pumps are mostly used to handle large flows of saturated feed water at high temperature and pressure. Compared to thermal power stations, the pumps for nuclear plants may have to handle higher flows at moderate heads.

The pump hydraulics is mainly decided by the head, capacity, speed and suction conditions. The required hydraulics is achieved by providing suitable clearances and impeller dimensions.

Table 1 Rajasthan atomic power plant at Kota

Design Parameters for Boiler Feed Pumps	
Number of boilers	8
Total steam output	1162 t/h
Steam pressure at drum	40 kg/cm^{2}g
Steam temperature at drum	250°C
Moisture in steam	0.25%, max
Feedwater flow	1072.5 t/h
Feedwater temperature	171°C
Reheater condensate return	89.5 t/h
Deaerator height above pump cl	23.77 m

Margin in Head and Capacity

A proper margin in head and capacity should always be included over and above the conditions of service based on maximum flows to be encountered in operation to compensate for the effect of wear in service and as a precaution against under-estimating the maximum capability of the main unit or system pressure losses. As a general rule, an 8% margin to the maximum boiler capacity is considered in arriving at the rated capacity of the boiler feed pump installation. Over-liberal margins result in power wastage and excessive wear of regulating valves especially in motor-driven constant-speed pumps. The margin can be added to the desired capacity or to the required total head, the net result on the head-capacity curve is essentially the same. But adding margin to the total head has the advantage of keeping design capacities nearer to the best efficiency point of the pump curve; otherwise the pump will be running constantly at a lower efficiency and in a range where serious pulsations and hydraulic surges can take place. In order to overcome the above adverse effects of oversizing, the margin in head or capacity of boiler feed pumps has been reduced to as little as 4% in the latest utility power plants.

The drum pressure to be used in calculating the rated pump conditions need not incorporate the 6% over-pressure setting of the safety valves. The average boiler feed pump head-capacity curve has a rise of 15% to 20% from its design head to shut off, the steepest portions occurring near the rated capacity. Even if the pump had to feed water into the boiler under the adverse conditions of all the safety valves blowing, it would still be able to maintain a very respectable rate of flow into the boiler and protect it against mishap. Therefore, the pump should not be oversized as it leads to much wasted power and requires excessive pump and valve maintenance due to throttle wear.

The specification of the boiler feed pumps intalled at Rajasthan Atomic Power Plant at Kota shows a margin of 10.6% on the rated capacity which is little on higher side (Table 2).

Pump Speed

The boiler feed pumps are very large pumps and may consume power as high as 2% of the rated capacity of the turbo-generator. Therefore, very high efficiencies are expected from these types of pumps. Considering that pump efficiency increases as specific speed increases, the boiler feed pumps are designed with specific speeds in the range of 2000 to 2500 (FPS units). However, the pump speed will also depend upon the type of driver used and its speed.

In the case of multistage pumps for high heads, the desire to increase the specific speed leads to a reduction of the head per stage and hence to an increase in the number of stages required. This in turn will result in longer shaft spans, greater static deflection, and thus lower reliability. The design for slightly higher efficiency under these circumstances clearly works against the user.

Table 2 Rajasthan atomic power plant at Kota

Specifications: Boiler Feed Pumps	
Number of pumps	2 × 100% duty
Rated capacity	19985 lpm
Total head	500 m
Speed	3000 rpm
Type of pump	Multi-stage constant speed centrifugal pumps
Number of stages	3
Manufacturer	Bingham pump Co Ltd, Canada
Margin on capacity	10.6%
Head per stage	166.7 m
Recirculation bypass capacity	6.6% of rated capacity
Specific speed	2000 FPS units
Number of impeller vanes	
1st stage = 5 original, 7 after modifications	
2nd stage = 7	
3rd stage = 7	

In order to obtain high pump reliability, high speed pumps are used which can give higher heads per stage and develop the required pressure with lesser number of stages. Here the speed is limited by the capabilities of its driver and availability of NPSH of the installation. For a given specific speed and fixed head per stage, the pump speed must decrease with increasing capacity.

When integrated properly into the station heat balance, the use of auxiliary steam turbines can provide significant savings in overall heat rate and for this reason there is a strong trend towards their use as boiler feed pump drives in thermal power plants. However, in nuclear power plants, electric motors are preferred as pump drivers to reduce initial cost and to avoid complicated layout. The steam turbine is inherently a variable speed machine and does not require the use of a variable speed drive such as a hydraulic coupling or the magnetic drive. The most important attraction of the use of steam turbine is increased net power output of the station by eliminating the electric power required by the boiler feed pumps.

The hydraulic design data of the boiler feed pumps used at Rajasthan Atomic Power Plant at Kota are given in Table 2. The data confirms to the prevailing practice of boiler feed pump design.

Suction Conditions

There is a danger of flashing in the pump suction whenever the load on the steam turbine is dropped suddenly and there is a pressure drop in the bleed line from turbine to deaerator. The water in the storage space of the deaerator and pump suction line is at a higher temperature than that corresponding to the reduced pressure. Violent vaporization of water will take place until the pressure-temperature equilibrium is established again.

A sudden 'dynamic' pressure drop may occur in the suction line when the discharge valve is suddenly opened to meet increased demand and again creating the conditions of flashing. Therefore, NPSH requirements for boiler feed pumps are higher than required as per cavitation constant sigma to prevent vapour binding as a result of sudden reduction of electric load, or sudden increase of pump capacity.

Recirculation Capacity

There is an increase in enthalpy of feed water from suction to discharge manifesting as temperature rise. There can be overheating of pumps when the capacity is reduced below a safe limit as enthalpy increase is absorbed by a smaller quantity of water. The minimum flow required through the pump is calculated for each of the following situations and final decision is then based on the greatest of these minimua.

(i) increase in radial thrust at reduced flows,

(ii) temperature increase of water from suction to discharge,

(iii) internal recirculation in the pump suction or discharge or both at reduced flows,

(iv) excessive power requirement at reduced flows overloading the driver, and

(v) air or vapour lock at reduced flows.

The amount of water bypassed varies from 5% to 10% of the normal capacity. If this limit is exceeded and bypass flow is connected back to the pump suction, flashing can occur as higher temperature water is mixed at the pump suction. The larger sizes of boiler feed pumps and the higher design pressures have led to very significant values of recirculation bypass, requiring large and expensive valves and introducing appreciable disturbances when surges arise from the sudden increase or decrease in flow as the bypass controls open or close the recirculation. Dual bypass control valves and orifices working in parallel can be used. As the feed water flow decreases, the first of the two valves opens and with further reduction in flow the second valve opens. The closing of valves as demand increases follows the same pattern. The feed water in the bypass is returned to the deaerator and there will be saving in the supply of steam for deaeration.

Impeller Geometry

In addition to sudden changes in pump speed due to abrupt variations in electric loads or sudden increase of pump capacity, there are chances of flow separation from the vanes in case of operation at reduced flows. These transient conditions of operation can cause hydraulic surging (self-excited vibrations) leading to unstable operation, sharp and jerky pressure rise and damage similar to cavitation. The design of impeller to prevent hydraulic surging should include rounding off of inlet vane edge, increase in number of vanes and highly backward-curved vanes. The increase in a number of vanes ensures favourable cross-sectional shape but reduces the flow area affecting the impeller flow capacity. Normally seven vanes are used in boiler feed pump design.

In boiler feed pumps where low NPSH requirements are important to restrict the height of deaerator installation within reasonable limits, the entrance velocities are kept low. It is customary to use 'exaggerated' impeller suction eye to reduce entrance velocities. This, however, can lead to internal recirculation of flow within the pump when operated at reduced flows.

OPERATION AT REDUCED FLOWS

Operation of centrifugal pumps at reduced flows can lead to a number of unfavourable results which must be anticipated or circumvented.

(i) Obviously, the pump operates at less than its best efficiency.

(ii) Particularly if the pump is of single-volute design, it will develop a radial thrust which increases the load on its radial bearings. A pump that will operate at such flows must be designed to accept this higher bearing load.

(iii) As the capacity is reduced, the temperature rise through the pump increases. To avoid exceeding permissible limits, a minimum flow bypass must be provided.

(iv) At certain flows below that of best efficiency, centrifugal pumps are subject to internal recirculation both in the suction and discharge areas of the impeller. This can cause hydraulic surging and damage similar to cavitation damage.

INTERNAL RECIRCULATION

The phenomenon of internal recirculation and prediction of and control of magnitude of flows at which this occurs have been researched only recently.

The constant competitive pressure on manufacturers to build pumps with lower and lower NPSH values forced them to design impellers with large eye diameters. While the resulting lower velocities have little effect on the pump performance at or near its best efficiency capacity, they may lead to hydraulic surges, noisy operation and premature wear at low capacities. These symptoms are caused by internal recirculation at the suction. The flow at the outer eye diameter tends to reverse itself at low flows and causes excessive turbulence through the formation of a vortex.

The recirculation of impeller discharge is somewhat similar to the recirculation at the suction. It is engendered by the desire to improve the value of the pump efficiency by increasing the head coefficient of the impeller. This recirculation also creates hydraulic surges and local cavitation at the impeller tips. In multistage pumps, the suction and discharge recirculation are generally limited to the first stage and the subsequent stages have a greater margin between the prevailing pressure and liquid vapour pressure and the effect is either minimised or disappears entirely.

Antistall Rings

Certain procedures are available to the designer to minimize the effect of recirculation and to reduce the percentage of best efficiency capacity at which significant suction recirculation starts. These procedures generally increse the minimum required NPSH. Therefore, if the pump requires frequent and extended operation at reduced flows, it is better to select a pump with proper NPSH so that it will not suffer from the effects of recirculation.

For the existing pumps, certain modifications are possible. The simplest method is to install an antistall ring in the suction eye of the pump by shring-fitting to reduce the suction eye area and thereby the tendency for recirculation.

Minimum Flow Bypass

A centrifugal pump can be designed only for one capacity at a given head and speed, the geometry of the impeller can be ideal for the rated conditions; for all other flows below the rated capacity, the distribution of velocities and flow angles are distorted. The required NPSH does not follow the 'square of the flow' rule but decreases at a lesser rate.

The incipient internal recirculation takes place at reduced flow which depends upon each impeller design. It is not possible to calculate or measure the exact value of NPSH required to prevent this cavitation. It has however been established that an impeller with very low NPSH and extremely oversized eye diameter develops an internal recirculation at flows considerably higher than impeller with moderate NPSH requirements. Therefore, the minimum flow bypass line should be designed according to the tendency of the pump to develop internal recirculation.

On installations already existing, the minimum flow bypass capacity may have to be increased if internal recirculation occurs. This may be required in addition to the installation of antistall rings in the pump suction eye.

Cold Water Injection into Pump Suction

Providing sufficient NPSH for the boiler feed pump by raising the height of deaerator or other means can pose practical difficulties, which can be overcome by injecting cold water from a suitable point in the feed water cycle when the pump is operating at reduced capacities. Injection of cold water into the pump suction results in the subcooling and available NPSH is artificially increased. The amount of cold water to be injected can be calculated from the amount of subcooling (temperature depression) required by energy balance.

In the case of high-pressure boilers where the possible effect of oxygen contamination may be quite severe, the colder water injection has to be tapped from downstream side after the deaerator. A heat exchanger is installed in the suction line of the pump to subcool the water and cooling medium employed can be condensate from the inlet to the deaerator. In order to avoid complication of the subcooling scheme, the cold injection can be directly taken from the condenser hotwell where deaeration of water is reasonably well and bypassing the deaerator by a small fraction (6% to 8%) of feed water may not have serious effects.

CASE STUDY OF RAPP – I

The pump specifications for the Rajasthan Atomic Power Plant (RAPP) are given in Table 2. These pumps together with other nuclear pumps were supplied by M/s Bingham Pump Co. Ltd., Canada, who had especially developed these pumps and elaborate testing was not carried out by the manufacturer. Subsequent to first commissioning and start-up, high vibrations and noise were noticed on the boiler feed pump. The commissioning tests were carried out at about 30 MW against the rated capacity of the unit of 220 MW as higher connected load was not available in the grid. The pump was dismantled for inspection to find out the reasons for vibrations. Some damage similar to cavitation was found on the first stage impeller wearing ring. The pump was again run installing the spare rotating assembly, but high vibrations persisted. The second pump also showed high vibrations.

The shaft mounted oil pump was replaced. Foundation design and quality of grouting were again checked. Detailed alignment checks were made. The individual motors were checked for smooth running.

As per the practice of the pump manufacturer, individual impellers and pump shafts were dynamically balanced separately without checking the balance of final rotor assembly. Therefore, the rotor assembly was taken to Walchannagar Industries for balance check as balancing machine was not available in the adjoining areas. The checks showed the imbalance within limits of international practice.

It was therefore concluded that the hydraulic design of the pump installation was insufficient for operation at reduced flow. Due to non-availability of grid capacity, it was not possible to check the performance of the pumps at higher loads. Vibration data were analysed in consultation with the manufacturer. As the pumps were the first of their model, elaborate test data were not available on hydraulic performance. The manufacturer suggested certain modifications based on their experience on similar pumps in the USA and Canada. It was a typical case of internal recirculation at reduced flows with consequent noise and vibrations.

(i) The profile of the volute sections of the first stage impeller was checked as per templates supplied by the manufacturer. Some grinding was necessary.

(ii) The discharge vane tips were chamferred to change the discharge angle.

(iii) An antistall ring was installed in the suction eye. The rotor assembly was mounted on a lathe to machine the internal diameter of the suction eye of first stage impeller to suit the OD of antistall ring. The impeller was uniformly preheated to 150°C and ring was shrunk-fit followed by dynamic check of the assembly.

(iv) The capacity of the bypass was increased from 22 lps to 63 lps by replacing the orifice block and control valves.

(v) The pump was tested after the above modifications but there was no appreciable improvement in the vibration level. Vibration data showed predominant vibrations at 5 N harmonics.

(vi) The first stage impeller with five vanes was replaced with one with seven vanes. (Number of vanes in the second and third stages were seven each.)

(vii) The capacity of the recirculation line was further increased to 145 lps by changing the pipeline from 75 NB to 200 NB. An orifice of 94 mm diameter was provided in the horizontal run of the bypass line near the deaerator. The size of the orifice was restricted by the maximum flow velocity permissible (4.88 m/s).

The boiler feed pumps were commissioned after carrying out all the above modifications. The pumps have been working smoothly without vibrations and caviation.

CASE V

Hydro Power Plants: Great Ecological Havocs

The utilization of energy for the performance of tasks has always been integral with man's standard of living. Different branches of industry, agriculture, transport, commercial activities, domestic comforts, entertainment, all depend upon ample supply of power. But power production is associated with disturbing the ecological balance and environmental pollution. The thermal power plants eject large quantities of solid and gaseous impurities and thermal discharges into the atmosphere, adversely affecting ecological conditions in the vicinity of the plant. Due to harmful effects of radioisotopes on living organisms, the handling of nuclear fuels and wastes requires utmost caution. The nuclear power plants pose great dangers to human life in the case of accidents.

Hydropower plants are considered the cleanest and least harmful of power plants because there is no ejection of toxic or radioactive materials or thermal discharge to the environments. However, there is a need for storage of large amounts of water behind the dams requiring submergence of vast areas of land under water. This results in large transfer of population and uprooting and rehabilitation of villages, towns and cities. The flow of water through the rivers downstream is reduced causing water pollution. The ecological balance can be much disturbed in some cases that hydropower plants can cause ecological havocs. These plants can be very dangerous hidden monsters. The damage caused by these plants have moral and social significance but the economical impact in terms of money lost has far reaching consequences. An attempt has been made to estimate the economical damages which can be caused by large hydropower plants with upstream reservoirs.

SLOW POISONING

During my school days, I happened to attend one election meeting in Punjab. The speaker addressing the peasants informed them that the water they were getting in the irrigation canals was no good for cropping as the same had been diluted by extracting energy at the Bhakra Hydroelectric Plant. That election gimmick amused me very much. After so many years, I am compelled to take his words seriously. The fertility of Indian fields especially plains of northern part of India depends upon the alluvial soil and nutrients brought along by rivers from the mountains and picked up while flowing over hill slopes. Through all centuries, the fertility of these fields has been maintained by the feed of suspended nutrients by our rivers and irrigation canal system. For the construction of large power plants, these rivers are dammed for storage of water to ensure a regular flow with static head created. In the process, the water is decanted of most part of the suspended nutrients as the latter settles down as silt in the reservoirs by sedimentation. The fields are devoid of the natural fertilization process as these do not get the nutrients through irrigation water. The land can slowly become completely barren. It may be a very slow process taking hundreds of years like the slow lead poisoning of the Romans by the lead utensils used by them for drinking wine. If not prevented, the denial of natural fertilization can lead to very serious consequences over the years resulting in loss of fertile land for agriculture and other allied economic activities. This natural process has been the main source of flourishing of rich Indian civilization and agricultural society of Indus-Gangetic plains. By damming the rivers for the construction of hydro power plants, we may be working against a very important natural process of land fertilization.

The estimation of economical damages which may be caused by the loss of land fertility due the construction of a hydro power plant is a very complicated problem. The damage can be determined on the basis of the loss of crop due to exponential decrease in land fertility or by the additional expenditure required to prevent the loss of land fertility by the use of synthetic chemical fertilizers. The manufacture of these synthetic fertilizers consumes large quantities of useful natural resources and the energy spent in their manufacture may be much more than obtained from a corresponding hydro power plant. The long-term harmful effects of using the chemical fertilizers on soil structure and fertility are still not fully understood.

The damage to the agriculture due to this slow poisoning causing the loss of land fertility can be expressed as follows:

$$D_{sp} = \sum^{n} [D_w . e^{B_w t} + D_p . e^{B_p t} + D_a . e^{B_a t} + A_f . e^{B_f t}]$$

where,

D_{sp} = Damage due to slow poisoning of natural fertility loss.

D_w & B_w = Coefficients of damage due to withdrawal of land due to loss of land fertility.

D_p & B_p = Coefficients of damage due to loss of productivity in agriculture.

D_a & B_a = Coefficients of damage to animal breeding due to loss of fodder.

A_f & B_f = Coefficients of expenditure on increased use of chemical fertilizers.

n = Time under consideration.

LAND GRAB

The hydro power plants have a very low material utilization efficiency based on Einstein's Equation, $E = mc^2$. There is a need for storage of large amount of water behind the dams requiring submergence of vast areas of land under water. It takes 700×10^6 tonnes of water flow through a turbine of 1 GW capacity to generate 120×10^{12}J of energy. The intrinsic energy of this mass of water is 630×10^{26} J. Hence, the material utilization efficiency is $(120 \times 10^{12}/630 \times 10^{26}) \times 100\% = 0.19 \times 10^{-12}$ per cent only. Hydro power plants have the lowest material utilization efficiency compared to 10^{-2}% efficiency of nuclear power plants.

The storage of large quantity of water results in large transfer of population and uprooting and rehabiliation of villages, towns and cities. The local ecological conditions are adversely affected by the water reservoir behind the dams which effect the local climate, water pressure on the land and surrounding terrain enclosing the reservoir and water level of the underground water. This can lead to serious waterlogging problem affecting the crops, plants and animal lives in large surrounding areas.

In addition to moral and social consequences, the economical problems can be very serious for the affected population as echoed by the hill people against Tehri Dam in Uttarakhand. The ecological balance is so much disturbed in some cases that Silent Valley Project in Kerala had to be abandoned on this account. The seismic characteristic of the location can be completely changed by the hydrostatic pressure of the huge water reservoir created.

The damage due to land grab for submergence underwater may be estimated as follows:

$$D_{lg} = \sum^{n} (D_{agr} + D_{for} + D_m + D_r + A_r)$$

where,

D_{lg} = damages due to land grab,

D_{agr} = damage to agriculture,

D_{for} = damage of forestry,

D_m = damage to mining for non-availability due to submergence under water,

D_r = damage to fish farming,

A_r = expenditure for rehabiliation of population uprooted.

where,

$D_{agr} = D_w + D_p + D_a$

D_w = damage due to loss of availability of land for agriculture.

D_p = damage to agriculture productivity in the surrounding region due to waterlogging.

D_a = damage to the animal breeding due to loss of pastures and fodder.

Similar expressions can be developed for damage to forestry.

STARVATION

The flow of water through the rivers downstream is reduced due to construction of dams, causing water pollution. The capacity of rivers to carry the garbage of the cities and the towns and the effluents of the industries situated on their banks is greatly reduced. Similarly, the capacity of the thermal power and nuclear power plants and other industries located on the river banks is very much reduced for want of water. The reduced flow of water in the rivers results in the growth of blue algae, reproduction of epidemic bacteria, extinction of water meadows and salination of soil. In a nutshell, the socio-economic life of the people can be completely shattered.

It is technically feasible to construct a dam across the Straits of Gibralter and a super-high capacity hydro electric power house to feed whole of Europe with low-cost electric power. But this can lead to lowering of water level in the Mediterranean Sea causing abandonment of a number of existing ports and disturbing the socio-economic life of the South Europe; reduced water pressure can increase volcanic activity and disturb the climate; increase in salinity can adversely affect the sea flora and fauna.

The estimation of various monetary losses can be a complex problem requiring the concerted efforts of economists, technologists and other experts dealing with national economy. The loss due to water pollution caused by reduced water flow in the rivers can be estimated from the additional expenditure required to take up preventive measures such as construction of waste water treatment plants for the effluents from the industries and municipalities before discharging into the rivers. The losses can also be summed up from the damages caused to other water consumers such as power plants, other industrial enterprises, agriculture, fishery, etc. by the use of contaminated water. The calculations can be very complicated requiring the use of special nomograms.

The loss of industrial sector can be calculated as follows:

$$D_1 = D_i^1 + D_i^2 + D_i^3 + D_i^4 + D_i^5$$

where,

D_i^1 = losses due to unscheduled equipment breakdown and repairs there of.

D_i^2 = losses due reduced output,

D_i^3 = losses due to poor quality of products,

D_i^4 = losses due to decrease in service life of plant and machinery,

D_i^5 = losses due to rejection of products.

The polluted water has also long-term adverse effects on agriculture, forestry and fishery.

$$D'_{agr} = D'_w + D'_p + D'_a$$

where,

D'_{agr} = loss to agriculture,

D'_{w} = loss due to withdrawal of land from agriculture caused by pollution,

D'_{p} = loss due to fall in agricultural productivity,

D'_{a} = loss of animal breeding.

Similarly,

$$D_f = D_{f1} + D_{f2} + D_{f3} + D_{f4} + D_{f5}$$

where,

D_f = loss to fishery by polluted water,

D_{f1} = loss due to death of mature fish,

D_{f2} = loss due to reduced reproduction,

D_{f3} = loss due to perishing of nutritional organisms,

D_{f4} = loss due to spoiling of spawning areas,

D_{f5} = loss due to impaired quality of fish products.

Similarly, loss to forestry due to polluted water,

$$D_{fo} = D_{fo1} + D_{fo2} + D_{fo3}$$

where,

D_{fo1} = loss due to inhibited plant growth,

D_{fo2} = loss due to bad quality of commercial timber,

D_{fo3} = extra expenditure for cleaning and renewal of forests

Similarly, damage to public health D_{ph} includes expenditure on curing of diseases caused by the consumption of unhealthy water and loss of working hours due to the diseases, etc.

MECHANICAL EXPLOSION

The huge amount of water stored at the back of the dam has a great potential to cause death and destruction to hundred and thousands of cities; towns and villages which are situated in the vicinity of the hydro power plants. It can completely wipe out the civilization of a region from the surface of the earth in case of any damage to the dam due to earthquake or material failure or bombardment by the enemy. The mechanical explosion releasing suddenly millions of tonnes of water gushing out with very high momentum can be more devastating than much feared nuclear bombs. It must also be remembered that 90% of earthquakes take place in areas without previous seismic history. The economic loss due to sudden failure of a dam is tremendous and cannot be visualized.

CONCLUSION

It is clear from the foregoing discussions that the so called cleanest and least harmful hydro power plants are the most polluting and dangerous power plants. These should be rightly feared as their destruction power can be much more than nuclear bombs. Their effects on the socio-economic life of the people have far-reaching consequences. These can reduce the prosperous and fertile lands to barren wastelands not fit for any cropping. All the factors given in above must be properly weighed before planning a hydro power plant.

The hydro power, available free of cost in the form of flowing water is an important renewable source of energy. The large potential of hydro energy available from flowing water in the rivers and irrigation canals can be tapped by building run-off river and small hydro power plants without involving the construction of storage dams.

The adverse ecological effects and other dangers listed above may be avoided. The construction of dams requires huge investments and long gestation periods offsetting the economic edge of these types of plants over thermal plants. Hydro villages based on small hydro turbines can be very effective in bringing economic prosperity to the population living along the rivers and irrigation canals.

REFERENCES

1. Singal, R.K., Environmental Pollution by Power Industry—An Overall view, *Electrical India*, 15th February, 84.
2. Singal, R.K. An Alpine Hydro Village, *Electrical India*, 15th October, 1985.
3. Richter, Volkov and Pokrovosky, Thermal Power Plants and Environmental Control, Mir Publications, Moscow, 1985.
4. Venikov and Putyatin, *Introduction to Energy Technology*, Mir Publishers, Moscow,1981.

PART VI

OBJECTIVE QUESTIONS

1. FLUID FLOW

Tick (√) the right answer.

1. The expression $\frac{\partial \phi}{\partial t} + \int \frac{\partial p}{\rho} + \frac{1}{2}(\Delta\phi)^2 + gz$ = constant, represents
   (a) Steady flow energy equation
   (b) Unsteady irrotational Bernoulli's equation
   (c) Steady rotational Bernoulli's equation
   (d) Unsteady rotational Bernoulli's equation
2. Euler equation of turbine giving energy transfer per unit mass E_o (where U, V_w, V_r and V represent the peripheral, whirl, relative and absolute velocities respectively; suffixes 1 and 2 refer to the turbine inlet and outlet respectively) is given by
   (a) $E_o = U_1V_{w1} - U_2V_{w2}$ (b) $E_o = U_1V_{r1} - U_2V_{r2}$.
   (c) $E_o = U_1V_1 - U_2\,V_2$ (d) $E_o = V_1w_{w1} - V_2V_{w2}.V$
3. Which of the following assumptions are made for deriving Bernoulli's equation?
   (1) Flow is steady and incompressible.
   (2) Flow is unsteady and compressible.
   (3) Effect of friction is neglected and flow is along a stream line.
   (4) Effect of friction is taken into account and flow is along a stream line.

   Codes
   (a) (1) and (3) (b) (2) and (3)
   (c) (1) and (4) (d) (2) and (4)
4. The force of impingement of a jet on a vane increases if
   (a) The vane angle is increased.
   (b) The vane angle is decreased.
   (c) The pressure is reduced.
   (d) The vane is moved against the jet.
5. A streamline is a line
   (a) Which is along the path of the particle.
   (b) Which is always parallel to the main direction of flow.
   (c) Along which there is no flow.
   (d) On which tangent drawn at any point gives the direction of velocity.
6. The velocity potential of a velocity field is given by $\phi = x^2 - y^2$ + const. Its stream function will be given by
   (a) $-2xy$ + const. (b) $+2xy$ + const.
   (c) $-2xy + f(x)$ (d) $-2xy + f(y)$

7. The stream function in a 2-dimensional flow field is given by ϕ. The potential function is

   (a) $\dfrac{(x^2 + y^2)}{2}$; (b) $\dfrac{(x^2 - y^2)}{2}$;

   (c) xy (d) $x^2y + y^2x$.

8. A symmetrical stationary vane experiences a force 'F' of 100 N as shown when the mass flow rate of water over the vane is 5 kg/s with a velocity 'V' 20 m/s without friction. The angle α of the vane is

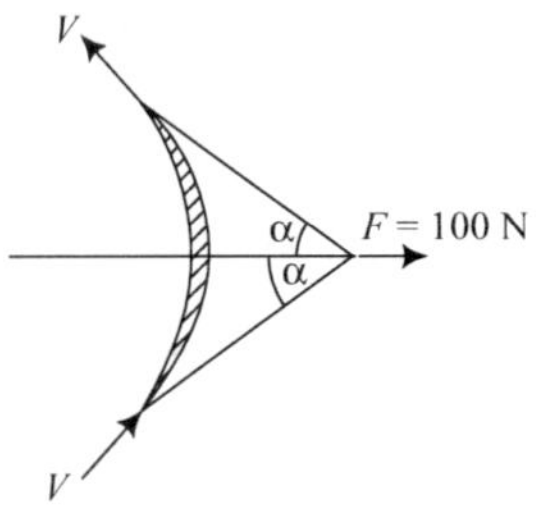

   (a) zero (b) 30°
   (c) 45° (d) 60°

9. In fluid machinery, the relationship between saturation temperature and pressure decides the process of:
   (a) flow separation (b) turbulent mixing
   (c) cavitation (d) water hammer.

10. The stream function is given by $(x^2 - y^2)$. The potential function of the flow will be
   (a) $2xy + f(x)$ (b) $2xy$ + const.
   (c) $2(x^2 - y^2)$ (d) $2xy + f(y)$

11. The expression $\left(\rho + \rho gz + \dfrac{\rho V^2}{2}\right)$ commonly used to express Bernoulli's equation has units of
   (a) total energy per unit mass.
   (b) total energy per unit weight.
   (c) total energy per unit volume.
   (d) total energy per unit cross-sectional area of flow.

12. The differential form of continuity equation for two-dimensional flow of fluid may be written in the following form

$$\rho\frac{\partial u}{\partial x} + \rho\frac{\partial v}{\partial y} = 0$$

in which u and v are velocities in the x and y directions and ρ is the density. This is valid for:

(a) compressible, steady flow
(b) compressible, unsteady flow.
(c) incompressible, unsteady flow
(d) incompressible, steady flow.

13. A 2-dimensional flow in x-y plane is irrotational if

(a) $\dfrac{\partial u}{\partial x} = \dfrac{\partial v}{\partial y}$ (b) $\dfrac{\partial u}{\partial x} = \dfrac{\partial u}{\partial y}$

(c) $\dfrac{\partial v}{\partial x} = \dfrac{\partial u}{\partial y}$ (d) $\dfrac{\partial v}{\partial x} = \dfrac{\partial v}{\partial y}$

14. If ψ = stream function and ϕ = velocity potential, the incorrect relation for irrotational flow is,

(a) $\psi = xy$ (b) $\psi = A(x^2 - y^2)$

(c) $\phi = ur \cos\theta + \dfrac{u}{r}\cos\theta$ (d) $\phi = \left(r - \dfrac{2}{r}\right)\sin\theta$

15. Euler equation for water turbine is derived on the basis of:

(a) conservation of mass
(b) rate of change of linear momentum
(c) rate of change of angular momentum
(d) rate of change of velocity

16. Select the correct statements regarding 2-dimensional potential flow.

(1) Laplace equation for stream function must be satisfied.
(2) Laplace equation for velocity potential must be satisfied.
(3) Stream lines and equipotential lines are mutually perpendicular.
(4) Stream function and potential functions are not interchangeable.

Codes

(a) (1) and (4) (b) (2) and (4)
(c) (1), (2) and (3) (d) (2), (3) and (4)

17. The Euler equations of motion for the flow of an ideal fluid is derived considering the principal of conservation of:

(a) mass and fluid as incompressible and viscous.
(b) momentum and fluid as incompressible and viscous.
(c) momentum and fluid as incompressible and inviscid.
(d) energy and the fluid as incompressible and inviscid.

18. The continuity equation for a steady flow states that
    (a) the velocity field is continuous at all points in flow field.
    (b) the velocity is tangential to stream lines.
    (c) the stream function exists for steady flows.
    (d) the efflux rate of mass through the control surface is zero.
19. Bernaulli's equation represents the
    (a) forces at any point in the field and is obtained by integrating the momentum equation to viscous flows.
    (b) energies at any point in the flow field and is obtained by integrating the Euler equations.
    (c) momentum at any point in the flow field and is obtained by integrating the equation of continuity.
    (d) moment of momentum and is obtained by integrating the energy equation.
20. The streamlines and the lines of constant velocity potential in an inviscid rotational flow field form
    (a) parallel grid lines placed in accordance with their magnitude.
    (b) intersecting grid net with arbitrary orientation
    (c) an orthogonal grid system
    (d) none of the above.
21. A 2-dimensional fluid flow is described by velocity components

$$U = 5x^3 \text{ and } V = -15x^2y.$$

    The stream functions will be
    (a) 2 m^3/s (b) 5 m^3/s
    (c) 10 m^3/s (d) 20 m^3/s

Key

1. (b)	2. (a)	3. (a)	4. (d)	5. (d)	6. (a)	7. (b)
8. (d)	9. (c)	10. (b)	11. (c)	12. (d)	13. (c)	14. (d)
15. (c)	16. (c)	17. (c)	18. (d)	19. (b)	20. (c)	21. (c)

2. TURBINES

Tick (√) the right answer.

1. Two Pelton wheels A and B have the same specific speed and are working under the same head. Wheel A produces 400 kW at 1000 rpm. If *B* produces 200 kW, then its rpm is
   (a) 4000 (b) 2000
   (c) 1500 (d) 1250
2. Consider the following types of water turbines:
   (1) Bulb (2) Francis
   (3) Kaplan (4) Pelton
   The correct sequence of order in which the operating head decreases while developing the same power is
   (a) (4) – (2) – (3) – (1) (b) (3) – (4) – (1) – (2)
   (c) (2) – (1) – (4) – (3) (d) (1) – (3) – (2) – (4)
3. Consider the following energies associated with a Pelton turbine
   (1) Mechanical energy (2) Kinetic energy
   (3) Potential energy
   The correct sequence of energy conversion starting from the entry of fluid is:
   (a) (1) – (2) – (3) (b) (2) – (3) – (1)
   (c) (3) – (2) – (1) (d) (1) – (3) – (2)
4. Which one of the following graphs correctly represents the relations between head and specific speed for Kaplan and Francis turbines.

   (a)
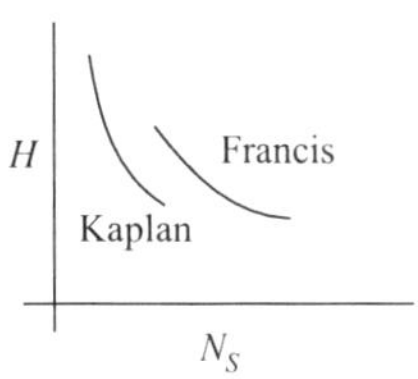

   (b)
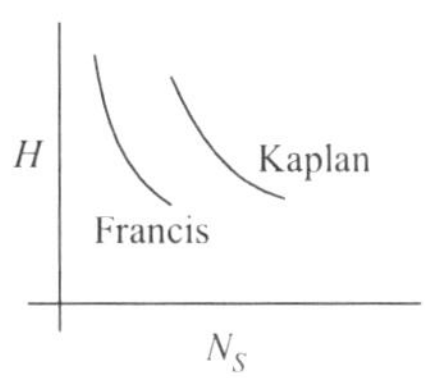

   (c)
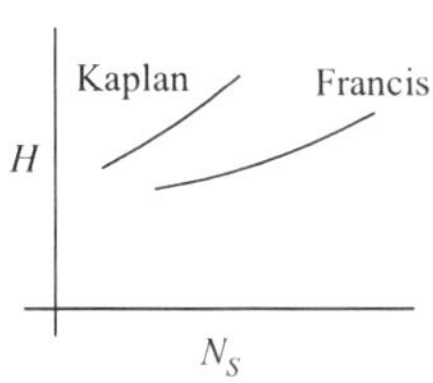

   (d)
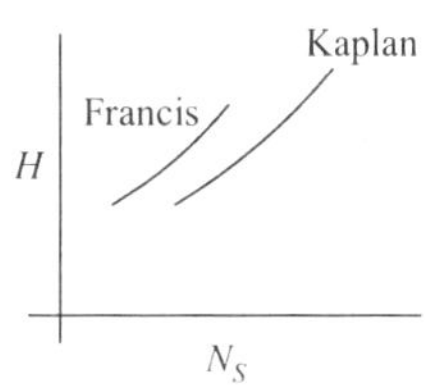

5. A draft tube is used in a reaction turbine
   (a) to guide water downstream without flashing.
   (b) to convert residual pressure energy into kinetic energy.

(c) to convert residual kinetic energy into pressure energy.
(d) to streamline the flow in the tail race.

6. In the phenomenon of cavitation, the characteristic fluid property involved is:
(a) surface tension (b) viscosity
(c) bulk modulus of elasticity (d) vapour pressure.

7. Which of the following statements are correct?
(1) Pelton wheel is a tangential flow impulse turbine.
(2) Francis turbine is an axial flow reaction turbine
(3) Kaplan turbine is a radial flow reaction turbine

Codes

(a) (1) and (3) (b) (1) alone
(c) (2) alone (d) (3) alone

8. Efficiency of Pelton wheel is maximum if the ratio of jet velocity to tangential velocity of the wheel is
(a) $\frac{1}{2}$ (b) 1
(c) 2 (d) 4

9. The maximum efficiency in the case of Pelton wheel is (where angle of deflection of the jet = 180 – β)
(a) $\frac{1-\cos\beta}{2}$ (b) $\frac{1+\cos\beta}{2}$
(c) $\frac{\cos\beta}{2}$ (d) $\frac{1+\cos\beta}{4}$

10. Match List-I with List-II and select the correct code.

List-I (Hydraulic Turbine)	**List-II (Application Area)**
A. Pelton Turbine	1. Low head, large discharge
B. Francis Turbine	2. Medium head, medium discharge
C. Kaplan Turbine	3. High head, low discharge

Code

	A	B	C
(a)	2	3	1
(b)	2	1	3
(c)	3	1	2
(d)	3	2	1

11. If H is the available head for a hydraulic turbine, the power, speed and discharge, respectively are proportional to:
    (a) $H^{1/2}, H^{1/2}, H^{3/2}$ (b) $H^{3/2}, H^{1/2}, H^{1/2}$
    (c) $H^{1/2}, H^{3/2}, H^{1/2}$ (d) $H^{3/2}, H^{1/2}, H$.
12. In utilizing scale models in the designing of turbo-machinery which of the following relationships must be satisfied?
    (a) $\dfrac{H}{ND^3}$ = Const. $\dfrac{Q}{N^2D^2}$ = Const.
    (b) $\dfrac{Q}{D^2\sqrt{H}}$ = Const; $\dfrac{H}{N^3D}$ = Const.
    (c) $\dfrac{P}{QH}$ = Const; $\dfrac{H}{N^2D^2}$ = Const.
    (d) $\dfrac{NQ^{\frac{1}{2}}}{H^{\frac{3}{2}}}$ = Const; $\dfrac{NP^{\frac{1}{2}}}{H^{\frac{3}{4}}}$ = Const.
13. Consider the specific speed ranges of the following types of turbines;
    (1) Francis (2) Kaplan (3) Pelton,
    The sequence of their specific speed in increasing order is;
    (a) (1), (2), (3) (b) (3), (1), (2)
    (c) (3), (2), (1) (d) (2), (3), (1)
14. The level of runner exit is 5 m above the tail race, and atmospheric pressure is 10.3 m. The pressure at the exit of runner for a divergent draft tube can be
    (a) 5 m; (b) 5.3 m;
    (c) 10 m; (d) 10.3 m;
15. The gross head on a turbine is 300 m. The length of penstock supplying water from reservoir to turbine is 400 m. The diameter of the penstock is 1 m and velocity of water through penstock is 5 m/s. If coefficient of fiction is 0.0098, the net head on the turbine would be, nearly
    (a) 310 m; (b) 295 m;
    (c) 200 m; (d) 150 m;
16. For a water turbine running at constant head and speed, the operating characteristic curves in the figure show that upto a certain discharge 'q' both output power and efficiency remain zero. The discharge 'q' is required to:

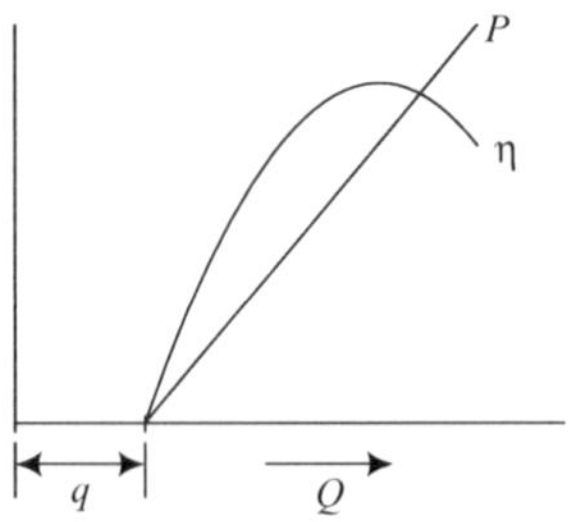

(a) overcome initial inertia
(b) overcome initial friction
(c) keep the hydraulic circuit full
(d) keep the turbine running at no load

17. The maximum number of jets generally employed in an impulse turbine without jet interference is
(a) 4
(b) 6
(c) 8
(d) 12

18. If the full-scale turbine is required to work under a head of 30 m and to run at 428 rpm, then a quarter scale turbine model tested under 10 m must run at:
(a) 143 rpm
(b) 341 rpm
(c) 428 rpm
(d) 988 rpm

19. The following pairs of formulas represent the specific speeds of turbine and pump respectively.
(a) $\frac{NQ^{\frac{1}{2}}}{H^{\frac{3}{4}}}$ and $\frac{NP^{\frac{1}{2}}}{H^{\frac{5}{4}}}$
(b) $\frac{NQ^{\frac{1}{2}}}{H^{\frac{3}{4}}}$ and $\frac{NP^{\frac{1}{2}}}{H^{\frac{3}{4}}}$
(c) $\frac{NP^{\frac{1}{2}}}{H^{\frac{3}{4}}}$ and $\frac{NQ^{\frac{1}{2}}}{H^{\frac{5}{4}}}$
(d) $\frac{NP^{\frac{1}{2}}}{H^{\frac{5}{4}}}$ and $\frac{NQ^{\frac{1}{2}}}{H^{\frac{3}{4}}}$

20. The cavitation number of a fluid machine is defined as

$$\sigma = \frac{p - p'}{\rho V^2 / 2}$$

where p = absolute pressure
ρ = density
V = free stream velocity

The symbol p' denotes

(a) static pressure of fluid
(b) dynamic pressure of fluid
(c) vapour pressure of fluid
(d) shear stress of fluid.

21. Which of the statements are correct regarding water turbine governor?

(1) It helps in starting and shutting down the turbo-unit.
(2) It controls the speed of turbine set to match it with the hydroelectric system.
(3) It sets the amount of load which a turbine unit has to carry.

Codes

(a) (1), (2) and (3) (b) (1) and (2)
(c) (2) and (3) (d) (1) and (3).

22. What is the correct sequence of turbines in increasing order of their specific speeds?

(1) Francis turbine
(2) Pelton wheel with two or more jets
(3) Pelton wheel with one jet
(4) Kaplan turbine

Codes

(a) (2), (3), (1), (4) (b) (3), (2), (1), (4)
(c) (2), (3), (4), (1) (d) (3), (2), (4), (1)

23. The correct sequence of hydraulic turbines in decreasing order of their specific speeds is:

(a) Pelton wheel, Francis turbine, Kaplan turbine
(b) Propellor turbine, Francis turbine, Pelton wheel
(c) Kaplan turbine, Pelton wheel, Francis turbine
(d) Francis turbine, Kaplan turbine, Pelton wheel.

24. A Pelton wheel is ideally suited for:

(a) High head and low discharge
(b) High head and high discharge
(c) Low head and low discharge
(d) Medium head and medium discharge

25. The following turbine is used in underwater power station

(a) Pelton turbine (b) Deriaz turbine
(c) Tubular turbine (d) Turgo-impulse turbine

26. The correct sequence in increasing order of specific speeds of the given turbines is

    (1) Kaplan (2) Pelton (3) Francis

    Code

    (a) (2), (3) and (1)
    (b) (2), (1), (3)
    (c) (3), (1), (2)
    (d) (1), (2), (3)

27. The following shape of draft tube will not improve the hydraulic efficiency of a turbine.

    (a) Straight cylindrical
    (b) Conical
    (c) Bell-mouthed
    (d) Bent tube

28. The specific speed of a hydraulic turbine is given by

    (a) $\dfrac{N\sqrt{P}}{H^{\frac{4}{5}}}$;
    (b) $\dfrac{N\sqrt{P}}{H^{\frac{5}{4}}}$;
    (c) $\dfrac{P\sqrt{N}}{H^{\frac{4}{5}}}$;
    (d) $\dfrac{P\sqrt{N}}{H^{\frac{5}{4}}}$.

29. The pressure acting on the runner of a reaction turbine varies from

    (a) more than atmospheric pressure to vacuum.
    (b) less than atmospheric pressure to zero gauge pressure.
    (c) atmospheric pressure to more than atmospheric pressure.
    (d) atmospheric pressure to vacuum.

30. Which statement is correct regarding an impulse turbine?

    (a) There is no pressure variation in flow over the buckets and the fluid fills the passage way between the buckets.
    (b) There is no pressure variation in flow over the buckets and the fluid does not fill the passage way between the buckets.
    (c) There is pressure drop in flow over the buckets and the fluid fills the passage way between the buckets.
    (d) There is pressure drop in flow over the bucket and the fluid does not fill the passage way between the buckets.

31. Kinematic similarity between the model and prototype is the similarity of

    (a) shape
    (b) discharge
    (c) stream
    (d) forces

32. In reaction turbines, the draft tube is used:

    (a) for the safety of the turbine
    (b) to convert kinetic energy of flow by a gradual expansion of the flow cross-section

(c) to destroy the undesirable eddies
(d) for none of the above purposes

33. The specific speed of a turbine is defined as the speed of a member of the same homogeneous series of such a size that it,
(a) delivers unit discharge at unit head.
(b) delivers unit discharge at unit power.
(c) delivers unit power at unit discharge.
(d) produces unit power under the unit head.

34. Match List-I with List-II and select the correct answer.

List-I	List-II
A. Pelton wheel (single jet)	1. Medium discharge, low head
B. Francis turbine	2. High discharge, low head
C. Kaplan turbine	3. Medium discharge, medium head
	4. Low discharge, high head

Codes

	A	B	C
(a)	1	2	3
(b)	1	3	4
(c)	4	1	3
(d)	4	3	2

35. The degree of reaction of a turbine is defined as the ratio of;
(a) static pressure drop to total energy transfer
(b) total energy transfer to static pressure drop.
(c) change of velocity energy across the turbine to the total energy transfer
(d) velocity energy to pressure energy.

36. The movable wicket gates of a reaction turbine are used to:
(a) control the flow of water passing through the turbine.
(b) control the pressure under which the turbine is working.
(c) strengthen the casing of the turbine.
(d) reduce the size of turbine.

37. Specific speed of a pump and specific speed of a turbine are:

(a) $\frac{N\sqrt{Q}}{H^{\frac{3}{4}}}$ and $\frac{N\sqrt{P}}{H^{\frac{5}{4}}}$; (b) $\frac{N\sqrt{Q}}{H^{\frac{3}{4}}}$ and $\frac{N\sqrt{P}}{H^{\frac{3}{4}}}$

(c) $\frac{N\sqrt{Q}}{H^{\frac{5}{4}}}$ and $\frac{N\sqrt{P}}{H^{\frac{5}{4}}}$; (d) $\frac{N\sqrt{Q}}{H^{\frac{5}{4}}}$ and $\frac{N\sqrt{P}}{H^{\frac{3}{4}}}$

38. Match List-I and List-II and select the correct statements.

List-I	List-II
A. Draft tube	1. Impulse turbine
B. Surging	2. Reciprocating pump
C. Air vessel	3. Reaction turbine
D. Nozzle	4. Centrifugal pump

Codes

	A	B	C	D
(a)	4	3	2	1
(b)	3	4	2	1
(c)	3	4	1	2
(d)	4	3	1	2

39. Match List-I and List-II and select the correct codes.

List-I (Turbines)	List-II (Characteristic)
A. Propeller	1. Inward flow reaction
B. Francis	2. Tangential flow impulse
C. Kaplan	3. Axial flow reaction with fixed vanes.
D. Pelton	4. Axial flow reaction with adjustable vanes

Codes

	A	B	C	D
(a)	2	4	1	3
(b)	3	4	1	2
(c)	2	1	4	3
(d)	3	1	4	2

40. Chances of occurrence of cavitation are high if the
    (a) local pressure becomes very high.
    (b) local temperature becomes high.
    (c) Thomas cavitation parameter exceeds a certain limit
    (d) local pressure falls below the vapour pressure.
41. Which of the following statements are false regarding the relation, specific speed, $N_S = \dfrac{N\sqrt{Q}}{H^{\frac{5}{4}}}$.
    (a) The relation is valid for both impulse and reaction turbines
    (b) The value of specific speed depends on the system of units.

(c) The value of specific speed for Pelton wheel will be lower than that for Kaplan turbine for any system of units.

(d) The dimensions of specific speed are (FLT^{-1}).

Key

1. (b)	2. (a)	3. (c)	4. (b)	5. (d)	6. (d)	7. (b)
8. (c)	9. (b)	10. (d)	11. (b)	12. (c)	13. (b)	14. (b)
15. (b)	16. (d)	17. (b)	18. (d)	19. (d)	20. (c)	21. (a)
22. (b)	23. (b)	24. (a)	25. (c)	26. (d)	27. (d)	28. (b)
29. (a)	30. (a)	31. (c)	32. (b)	33. (d)	34. (d)	35. (a)
36. (b)	37. (a)	38. (b)	39. (d)	40. (d)	41. (d)	

3. PUMPS

Tick (√) the right answer.

1. On the assumption that a double suction impeller is the equivalent of single suction impellers placed back to back, it is customary to base the specific speed of the double suction pump on,
   (a) One half of total capacity
   (b) Three-fourths of the total capacity
   (c) Full total capacity
   (d) Double the total capacity
2. Priming is necessary in
   (a) Centrifugal pumps to lift water from a greater depth.
   (b) Centrifugal pumps to remove air in the suction pipe and casing.
   (c) Hydraulic turbine to remove air in the turbine casing.
   (d) Hydraulic turbine to increase the speed of turbine and to generate more power.
3. A pump is installed at a height of 5 m above the water level in the sump. Frictional loss on the suction side is 0.6 m. If the atmospheric pressure is 10.3 m of water and vapour pressure head is 0.4 m (abs), the NPSH (Net Positive Suction Head) will be:
   (a) 3.7 m (b) 4 m
   (c) 4.3 m (d) 4.6 m
4. The pressure ratio of a pump and its $\frac{1}{4}$th scale model, if the ratio of heads is 5:1, will be
   (a) 100 (b) 3.2
   (c) 179 (d) 12.8
5. Consider the following statements regarding air vessels provided in reciprocating pump installations:
   (1) The air vessels are fitted both on suction and delivery sides.
   (2) The air vessels are fitted far from the pump cylinder.
   (3) The air vessels save energy by reducing the friction loss.

   Which of these statements are correct?
   (a) (1), (2) and (3) (b) (1) and (2)
   (c) (2) and (3) (d) (1) and (3)

6. Match List I with List II and select the correct answer using the codes given below:

List-I **Type of pumps**	**List-II** **Associated Features**
A. Centrifugal pump	1. Air vessel
B. Gear pump	2. Draft tube
C. Reciprocating pump	3. Guide vanes
D. Turbine pump	4. Rotary pump
	5. Rotor having blades

Codes

	A	B	C	D
(a)	4	2	5	3
(b)	5	4	1	2
(c)	4	2	3	1
(d)	5	4	1	3

7. Match List I with List II and select the correct answer using the codes given below.

List-I **(Industrial Needs)**	**List-II** **(Type of Pump)**
A. Combustible fluid to be pumped	1. Singlestage centrifugal.
B. High head but small discharge needed	2. Multistage centrifugal.
C. Low head but large discharge needed	3. Positive displacement.
D. High head and high discharge needed	4. Jet pump

Codes

	A	B	C	D
(a)	3	2	1	4
(b)	4	3	1	2
(c)	3	1	4	2
(d)	4	3	2	1

8. Fluid flow machines use the principle of either
   (i) supplying energy to fluid, or
   (ii) extracting energy from the fluid.

   Some fluid flow machines are a combination of both (i) and (ii). They are classified as:

   (a) compressors (b) hydraulic turbines
   (c) torque converters (d) windmills

9. A centrifugal pump driven by a directly coupled 3 kW motor of 1450 rpm speed, is proposed to be connected to another motor of 2900 rpm speed. The power of the motor should by:
   (a) 6 kW (b) 12 kW
   (c) 18 kW (d) 24 kW
10. A centrifugal pump gives maximum efficiency when its blades are
    (a) bent forward (b) bent backward
    (c) straight (d) wave shaped.
11. A pump running at 1000 rpm consumes 1 kW and generates head of 10 m of water. When it is operated at 2000 rpm, its power consumption and head generated would be:
    (a) 4 kW, 50 m of water (b) 6 kW, 20 m of water
    (c) 3 kW, 30 m of water (d) 8 kW, 40 m of water
12. The correct sequence of the centrifugal pump components through which the fluid flows is
    (a) Impeller, suction pipe, foot valve and strainer, delivery pipe.
    (b) Foot valve and strainer, suction pipe, impeller, delivery pipe.
    (c) Impeller, suction pipe, delivery pipe, foot valve and strainer.
    (d) Suction pipe, delivery pipe, impeller, foot valve and strainer.
13. In a fluid coupling, the torque transmitted is 50 kNm, when the speed of the driving and driven shaft is 900 rpm and 720 rpm respectively. The efficiency of the fluid coupling will be:
    (a) 20%; (b) 25%;
    (c) 80%; (d) 90%;
14. Which of the following statements are correct regarding fluid coupling?
    (1) Efficiency increases with increase of speed ratio.
    (2) Neglecting friction, the output torque is equal to input torque.
    (3) At the same input speed, higher slip requires higher input torque.

    Codes

    (a) (1), (2) and (3) (b) (1) and (2)
    (c) (2) and (3) (d) (1) and (3)
15. A hydraulic coupling transmits 1 kW of power at an input speed of 200 rpm, with a slip of 2%. If the input speed is changed to 400 rpm, the power transmitted with the same slip is:
    (a) 2 kW (b) 0.5 kW
    (c) 4 kW (d) 8 kW
16. If a reciprocating pump having a mechanical efficiency of 80% delivers water at the rate of 80 kg/s with a head of 30 m, the brake power of the pump is

(a) 29.4 kW; (b) 20.8 kW;
(c) 15.4 kW; (d) 10.8 kW.

17. Which of the following statements regarding a centrifugal pump are correct?
   (1) The monometric head is developed by the pump.
   (2) The suction pipe diameter is generally larger than discharge pipe diameter.
   (3) The suction pipe is provided with a foot valve and strainer.
   (4) The delivery pipe is provided with a foot valve and a strainer.

   Codes

   (a) (1), (2), (3) and (4) (b) (1) and (2)
   (c) (2) and (3) (d) (1) and (3)

18. Which of the following statements are correct regarding torque converter?
   (1) Its maximum efficiency is less than that of the fluid coupling.
   (2) It has two runners and a set of stationary vanes interposed between them.
   (3) It has two runners.
   (4) The ratio of secondary to primary torque is zero for zero value of angular velocity of secondary.

   Codes

   (a) (1) and (2) (b) (3) and (4)
   (c) (1) and (4) (d) (2) and (4)

19. Which of the following statements are correct regarding volute casing of a centrifugal pump?
   (1) Loss of head due to change in velocity is eliminated.
   (2) Efficiency of pump is increased.
   (3) Water from the periphery of the impeller is collected and transmitted to the delivery pipe at constant velocity.

   Codes

   (a) (1), (2) and (3) (b) (1) and (2)
   (c) (2) and (3) (d) (1) and (3)

20. Match List I with List II for a pump and select correct answer using the codes given below:

List-I (Outlet vane angle β_2)	**List-II (Performance curves)**
A. $\beta_2 < 90°$	
B. $\beta_2 = 90°$	
C. $\beta_3 > 90°$	

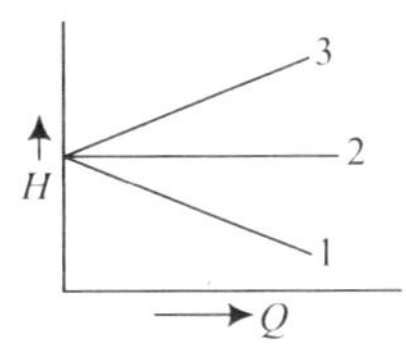

Codes

	A	B	C
(a)	1	2	3
(b)	1	3	2
(c)	2	1	3
(d)	3	2	1

21. Hydraulic ram is a pump which works on the principle of:
    (a) Water hammer (b) Centrifugal action
    (c) Reciprocating action (d) Hydraulic press

22. The data for the performance of a centrifugal pump:
    speed = 1200 rpm; flow rate = 30 l/s; head = 20 m; power = 5 kW
    If the speed is increased to 1500 rpm, the power will be nearly equal to
    (a) 6.5 kW (b) 8.7 kW
    (c) 9.8 kW (d) 10.9 kW

23. Which of the following pumps can be used to lift water through a suction head of 12 m from a well?
    (1) Centrifugal pump, single-stage
    (2) Centrifugal pump, multi-stage
    (3) Reciprocating pump
    (4) Jet pump
    Code
    (a) (2) alone (b) (1), (3) and (4)
    (c) (4) alone (d) (1) and (3)

24. The following purposes (given in code) are served by the volute casing of a centrifugal pump.
    (1) Increase in the efficiency of the pump.
    (2) Conversion of part of the pressure energy to velocity head.
    (3) Giving uniform flow of the fluid coming out of impeller.
    Code
    (a) (1) and (2) (b) (1) and (3)
    (c) (2) and (3) (d) (1), (2) and (3)

25. Which statements are correct regarding specific speed of a centrifugal pump?
    (1) Specific speed is defined as the speed of a genetically similar pump developing unit power under unit head.
    (2) At the same specific speed, the efficiency is greater with large capacity.

(3) The specific speed increases with increase of outer blade angle.

(4) The specific speed varies directly as the square root of the pump discharge.

Codes

(a) (1) and (2) (b) (2) and (4)

(c) (3) and (4) (d) (2) and (3)

26. Which are the correct statements regarding a torque converter?

(1) It has stationary set of blades in addition to the primary and secondary rotors.

(2) It can be used for multiplication of torques.

(3) The maximum efficiency of a converter is less than a fluid coupling.

(4) In a converter designed to give large increase of torque, the efficiency V/S speed ratio approaches unity.

Codes

(a) (1), (2), (3), (4) (b) (1), (2) and (3)

(c) (1), (2), (4) (d) (3) and (4)

27. In contrast to fluid coupling, torque converters are operated:

(a) while completely filled with liquid

(b) while partially filled with liquid

(c) without liquid

(d) while completely filled with air

28. A centrifugal pump is started with its delivery valve kept:

(a) fully open (b) fully closed

(c) partially open (d) 50% open

29. Select the correct statements. If pump NPSH is not satisfied, then

(1) It will not develop sufficient head to raise water

(2) Its efficiency will be low

(3) It will deliver very low discharge

(4) It will be cavitated.

Codes

(a) (1), (2), (3) (b) (1) and (4)

(c) (2), (3) and (4) (d) (1), (2), (3) and (4).

30. What are the beneficial effects of air vessel fitted to the delivery side of a reciprocating pump?

(1) Constant rate of discharge can be ensured.

(2) Power consumption can be reduced

(3) Discharge can be increased

(4) Constant velocity of piston can be ensured.

Codes

(a) (1) and (4) (b) (1) and (2)
(c) (2) and (4) (d) (1) and (3)

31. Choose the correct answers regarding place of cavitation in hydraulic machines.
    (1) exit of a pump
    (2) entry of pump
    (3) exit of a turbine.

    Codes

    (a) (1) and (2) (b) (1) and (3)
    (c) (1), (2) and (3) (d) (2) and (3)

32. For attaining a non-overloading characteristic in centrifugal pumps.
    (a) backward bent vanes are preferred over forward bent banes.
    (b) forward bent vanes are preferred over backward bent vanes.
    (c) forward bent vanes are preferred over vanes radial at outlet.
    (d) vanes radial at outlet are preferred over backward vanes.

33. Select the correct code regarding function of a volute casing of a centrifugal pump
    (1) To collect water from the periphery of the impeller and to transmit it to the delivery pipe at a constant velocity.
    (2) To increase discharge of the pump.
    (3) To increase the efficiency of the pump.
    (4) To reduce loss of head in discharge.

    Codes

    (a) (1), (2) and (3) (b) (2), (3) and (4)
    (c) (1), (3) and (4) (d) (1) and (2)

34. The specific speed of a hydraulic pump is the speed of geometrically similar pump against a unit head and
    (a) delivering unit quantity of water
    (b) consuming unit power
    (c) having unit velocity of flow
    (d) having unit radial velocity

Key

1. (c)	2. (b)	3. (a)	4. (c)	5. (d)	6. (d)	7. (b)
8. (c)	9. (d)	10. (b)	11. (d)	12. (b)	13. (c)	14. (a)
15. (a)	16. (a)	17. (c)	18. (a)	19. (a)	20. (a)	21. (a)
22. (c)	23. (c)	24. (a)	25. (a)	26. (a)	27. (a)	28. (c)
29. (b)	30. (b)	31. (d)	32. (a)	33. (c)	34. (a)	

4. PLANTS AND SYSTEMS

Tick (√) the right answer:

1. An accumulator is a device to store:
   (a) Sufficient quantity of liquid to compensate the change in discharge.
   (b) Sufficient energy to drive the machine when the normal energy source does not function.
   (c) Sufficient energy in case of machines which work intermittently to supplement the discharge from the normal source.
   (d) Liquid which otherwise would have gone waste.
2. A hydraulic press has a ram of 20 cm diameter and a plunger of 5 cm diameter. The force required at the plunger to lift a load of 16×10^4 N shall be:
   (a) 256×10^4 N (b) 64×10^4 N.
   (c) 4×10^4 N (d) 1×10^4 N
3. A surge tank provided on the penstock connected to a water turbine
   (1) Helps in reducing the water hammer
   (2) Stores extra water when not needed
   (3) Provides increased demand of water.
   Which of these statements are correct?
   (a) (1) and (3) (b) (2) and (3)
   (c) (1) and (2) (d) (1), (2) and (3).
4. The gross head available to a hydraulic power plant is 100 m. The utilized head in the runner of the hydraulic turbine is 72 m. If the hydraulic efficiency of the turbine is 90%, the pipe friction head is estimated to be:
   (a) 20 m (b) 18 m
   (c) 16.2 m (d) 1.8 m
5. A triangular dam of height h and base width b is filled upto its top with water as shown. The condition of stability is

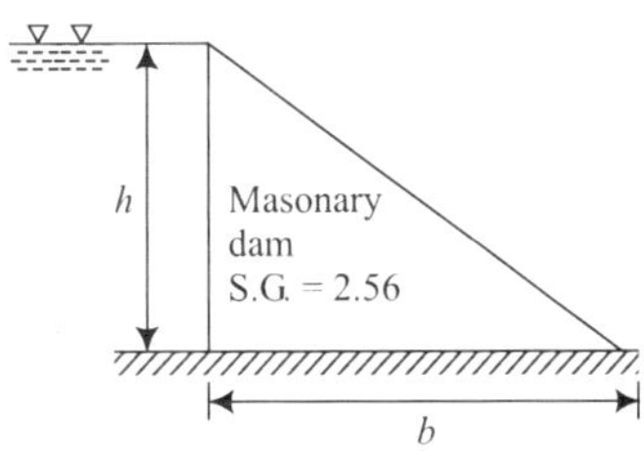

   (a) $b = h$
   (b) $b = 2.6h$
   (c) $b = \sqrt{3}\,h$
   (d) $b = 0.625h$
6. Of all power plants, hydel is most disadvantageous when one compares the
   (a) nearness to load centre
   (b) cost of energy resource
   (c) technical skill required
   (d) economics that determines the choice of plants

7. A hydraulic power station has the following major items in the hydraulic circuit:

(1) Draft tube;	(2) Runner;
(3) Guide wheel	(4) Penstock.

The correct sequence of these items in the direction of flow is;

(a) (4), (2), (3), (1), (5)	(b) (4), (3), (2), (5), (1)
(c) (1), (2), (3), (5), (4)	(d) (1), (3), (2), (4), (5)

8. Water hammer in penstocks takes place when
   (a) water is flowing with high velocity.
   (b) water is flowing with high pressure.
   (c) flowing water is suddenly brought to rest by closing a valve.
   (d) flowing water is brought to rest by gradually closing a valve.

Key

1. (c)	2. (d)	3. (d)	4. (a)	5. (b)	6. (d)	7. (b)
8. (c)						

Index